GUIDED WORKSHEETS

THINKING QUANTITATIVELY
COMMUNICATING WITH NUMBERS

SECOND EDITION

Eric Gaze

Bowdoin College

ISBN-13: 978-0-13-499639-4
ISBN-10: 0-13-499639-9

Guided Worksheets Table of Contents

WL 07.19.2019 1603

<u>Prior Knowledge Activation Worksheets</u>

Quantitative Reasoning / Quantitative Literacy Worksheet

Why go to college? What is the PURPOSE of a college education? List 3 specific purposes:

What is critical thinking? List 3 *characteristics* of critical thinking:

Summary

The purpose of this course is to give you the skills needed to make informed decisions when confronted with quantitative information in your personal, public, and professional lives. This introductory worksheet is simply meant to present you with quantitative information being presented as graphics like you will encounter in the popular press and in your courses.

At this point you are not expected to have the skills to answer the questions! They are meant to highlight how quantitative reasoning requires a different way of thinking and problem solving than is typically taught in traditional math courses. Try your best, work with your peers, and resolve to diligently work through the material in this course so that you will be among the few, the proud, the quantitatively literate :O)

A piece presented in the Bloomberg View explores the data regarding "How Americans Die."

https://www.bloomberg.com/graphics/dataview/how-americans-die/

Let's take a closer look.

1. On the first graph below use the statistics for 1968: 823.7, 967.3, and 1,118.5, in sentences.

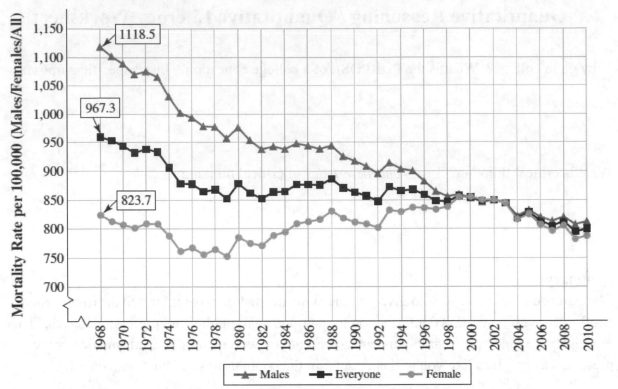

Data from: http://www.bloomberg.com/dataview/2014-04-17/how-americans-die.html

2. They tell us the overall rate "fell by about 17%" from 1968 to 2010, from 967.3 to 799.5.

a. Verify this and quantify the change for the rates for men and women in a similar fashion.

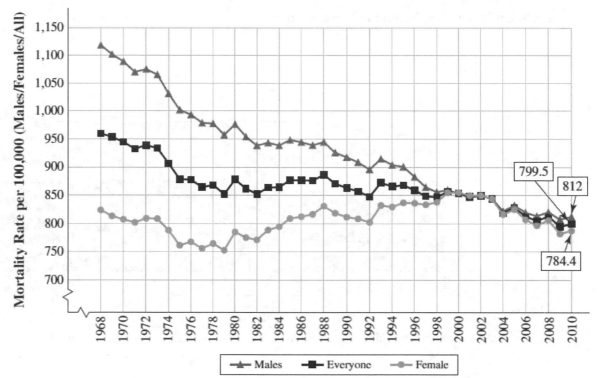

Data from: http://www.bloomberg.com/dataview/2014-04-17/how-americans-die.html

b. Why can we compare the 1970 and 2010 statistics, even though the population has increased over this period?

3. The slide associated with the first graph above says the decline in mortality rates stops in the mid 1990's and second slide with the graph below attributes this to the aging of the population.

a. What is the logic behind this argument?

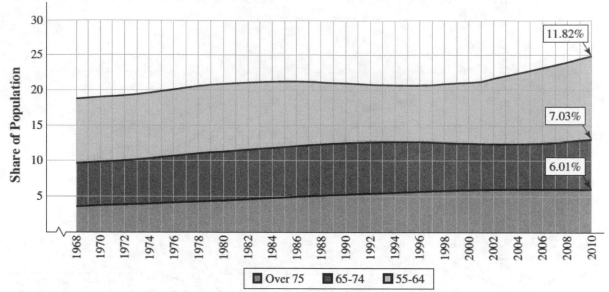

Data from: http://www.bloomberg.com/dataview/2014-04-17/how-americans-die.html

b. Interpret the 11.82% for 2010 in slide 2 and compare to the 25% on the vertical axis.

4. Looking at the graph below, which line stands out from the rest? What do you think accounts for this difference?

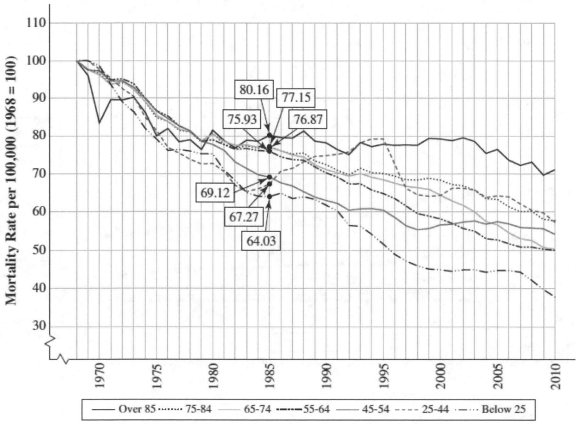

Data from: http://www.bloomberg.com/dataview/2014-04-17/how-americans-die.html

5. Interpret the statistics 80.16 and 64.03 for 1985 on graph in **#4**. Hint: Compare to the statistics shown here, and note that **in 1968** the mortality rate for over 85 was 19,598.5 per 100,000 and the rate for below 25 was 160.7 per 100,000.

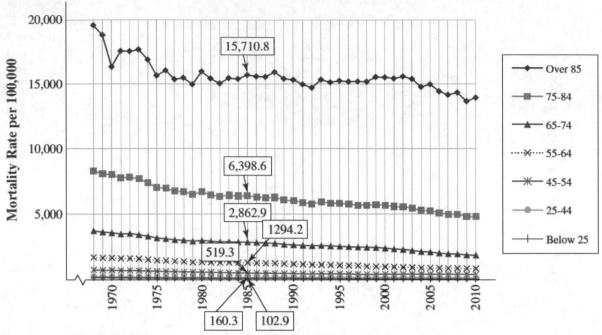

The ribbon provides nice shortcuts you can use to format your spreadsheet. When formatting numbers, it is important to remember that you are changing only the appearance of the number. Thus you may format a region as currency that will show only two decimal places, but the underlying number will have the full 16 decimal places. All of the underlying decimal places will be used in calculations not just the 2 showing. There are three common ways to format numbers as currency.

1. Highlight the numbers and click on the dollar sign button in the *Font* group.
2. Right-click on the highlighted region and choose *Format Cells*.
3. Click on the arrow next to the *Number* menu.

As you read, jot down notes and questions. At the end of this section's Guided Worksheets there is a space for **Reflection** *and* **Monitoring Your Understanding***, where you can try and answer your questions after completing the guided activities and homework.*

Notes & Questions

Key Terms

Formula	
Cell reference	
Active cell	
Formula bar	
Constant	
Parameter	

Summary

The screenshot below shows the *Sales Order Form* created in the e-text in the previous objective.

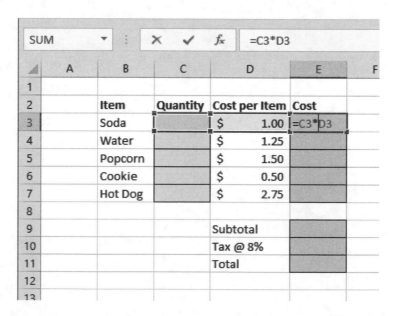

We have entered a **formula** in cell **E3** for the cost of a soda order using **cell references**. The symbol used for multiplication is *; the symbol for division it is /.

Cost of Soda = Quantity * Cost per Item

E3 = C3 * D3

The formula we type into cell **E3** is:

= C3 * D3

Notice that we start our formula with an equal sign, and when we type in **C3**, Excel highlights **C3**, and when we type in **D3**, Excel types in **D3**. **C3** and **D3** are called cell references. Our formula is referencing whatever values lie in those cells. Press **Enter** to enter this formula. You do not have to type in cell references, you can just click on the cells when typing in a formula. The output of this formula will be zero, represented as a dash in the *Accounting* format. Cell **C3** is blank and will be treated as zero in this computation.

When in typing in formulas there are some important things to remember:
1. The formula must start with an equal sign =.
2. You can click on a cell, **C3**, rather than typing in the cell reference.
3. The formula will show up in two places: in the **active cell** and also in the **formula bar** that is over the column headings. If you do not see the formula bar, go to the View menu and check the appropriate box in the Show menu to make it appear.

As you read, jot down notes and questions. At the end of this section's Guided Worksheets there is a space for **Reflection** *and* **Monitoring Your Understanding**, *where you can try and answer your questions after completing the guided activities and homework.*

Notes & Questions

Objective 3 – Fill Formulas

Key Terms

Fill	
Fill handle	

Summary

We need to enter the cost formulas for the other items: *Water, Popcorn, Cookie,* and *Hot Dog* in rows 4 – 7. Notice that these formulas are all very similar to the one we typed into cell **E3**.

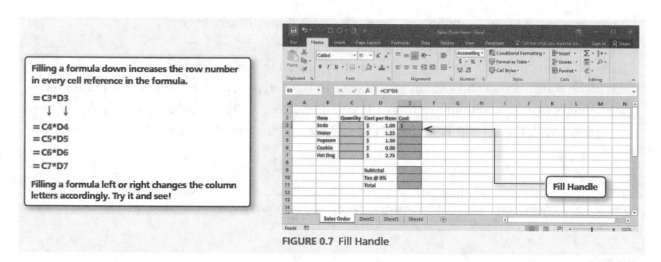

FIGURE 0.7 Fill Handle

Cost of Items Bought = Quantity * Cost per Item

E3 = C3 * D3

E4 = C4 * D4

Etc.

Rather than inputting this similar formula over and over we can **fill** the formula down by using the **fill handle**. You must first enter the formula in cell **E3** and press the Enter key so only the output of zero shows. Click back on cell **E3** to make it active, then click the fill handle on the lower right hand corner of the cell and drag down to cell **E7**.

As you read, jot down notes and questions. At the end of this section's Guided Worksheets there is a space for **Reflection** *and* **Monitoring Your Understanding**, *where you can try and answer your questions after completing the guided activities and homework.*

Notes & Questions

Guided Practice Activity #1 – Sales Order

We have updated the prices in the *Sales Order Form* shown below.

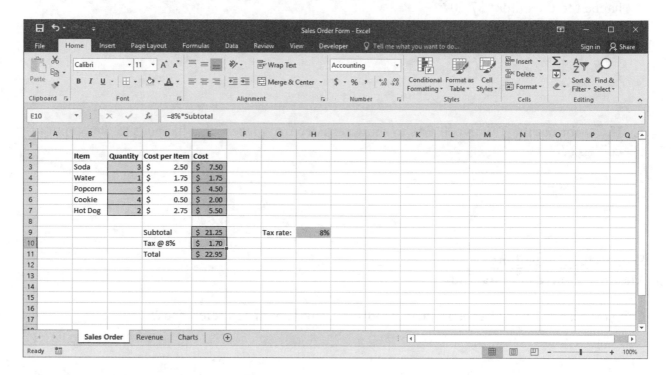

1. What formula is entered in cell **E3**?

2. How do you fill this formula down?

3. What happens to the cell references in the formula when you fill down?

4. What formula is in cell **E7**?

5. How do you format the numbers as currency?

6. What do you think would happen to the output $7.50 if you mistakenly format cell **E3** as a percentage by clicking on the **%** icon in the ribbon?

7. If you format cell **E3** to have 0 decimal places by clicking on the icons shown, what would happen to the output $7.50?

8. If you format the Tax in cell **E10** to show zero decimal places what happens to the output in cell **E11**?

Section 0.1 Reflections
Monitor Your Understanding

Can you describe what happens when you fill a formula with cell references?

Objective 1 – Using Built-In Functions

Key Terms

Built-in function	
Arguments	
Syntax	
Template	
Model	
Name box	

Summary

To compute the *Sub-Total*, we can sum of the item costs in cells **E3:E7**.

We can certainly enter the formula: = **E3+E4+E5+E6+E7**, but there is a faster way to accomplish this by using the built-in **SUM function** in cell **E9**:

$$= \text{SUM (E3:E7)}$$

Enter the formula for *Tax* in cell **E10** by taking 8% of the sub-total, and the formula for the *Total* in **E11** by adding the sub–total and the tax.

Excel has a huge suite of **built-in functions** for just about any computational task. There are five basic functions: **SUM**, **AVERAGE**, **MAX**, **MIN**, and **COUNT**. Note that **COUNT** only counts cells with numbers, and there are many other count functions! Each function must be typed in using proper **syntax** and proper **arguments**. The Insert Function button, *fx*, next to the formula bar can be a valuable help in typing in new functions. Every built-in function starts with an equal sign, like all formulas, and is followed by parentheses. The inputs, or **arguments**, for the function go in the parentheses and are separated by commas. A cell range, **E3:E7**, counts as just a single argument and is faster than entering: **E3,E4,E5,E6,E7**. We can see in the screenshot above that the arguments for the **SUM**(number1, [number2], …) are numbers. Arguments in square brackets are optional.

The *AutoSum* button in the *Editing* group is a quick way to enter the **SUM** function, or the other basic functions.

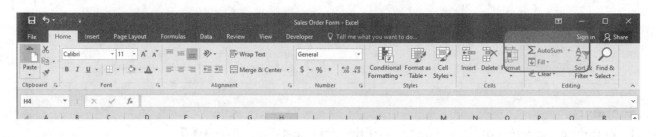

*As you read, jot down notes and questions. At the end of this section's Guided Worksheets there is a space for **Reflection** and **Monitoring Your Understanding**, where you can try and answer your questions after completing the guided activities and homework.*

Notes & Questions

What are the outputs of the following functions? Note cell **E3** has been formatted to show no decimal places.

SUM	▾	:	✕	✓	*fx*	=SUM(E3:E7)		
◢	A	B	C	D	E	F	G	H
1								
2		Item	Quantity	Cost per Item	Cost			
3		Soda	3	$ 2.50	$ 8			
4		Water	1	$ 1.75	$ 1.75			
5		Popcorn	3	$ 1.50	$ 4.50			
6		Cookie	4	$ 0.50	$ 2.00			
7		Hot Dog	2	$ 2.75	$ 5.50			
8								
9				Subtotal	=SUM(E3:E7)		Tax rate:	8%
10				Tax @ 8%	SUM(number1, [number2], ...)			
11				Total	$ 22.95			
12								

1. =AVERAGE(C3, C7)

2. =AVERAGE(C3:C7)

3. =SUM(C3:E3, E7)

4. =COUNT(B2:E11)

5. =MAX(C3:E11)

6. =MIN(C3, C7)

7. How do you name the cell **E9** "Subtotal"?

8. Naming a cell is like naming your dog, you really don't want to change the name! How do you un-name cell **E9**?

Section 0.2 Reflections
Monitor Your Understanding

Can you describe what a built-in function is?

Section 0.3 Charts

Objective 1 – Creating Charts

Key Terms

Chart	

Summary

We are now ready to go over the basics of creating a **chart** in Excel. We will use the Sales Order Form spreadsheet using the **Sheet 2** tab named Revenue and the data as shown.

◢	A	B	C	D	E	F
1						
2			Revenue per Item per Concert			
3			Concert 1	Concert 2	Concert 3	Totals
4		Soda	$ 56.00	$ 63.00	$ 53.00	$ 172.00
5		Water	$ 30.75	$ 45.00	$ 48.75	$ 124.50
6		Popcorn	$ 25.50	$ 30.00	$ 18.00	$ 73.50
7		Cookie	$ 36.00	$ 49.00	$ 44.50	$ 129.50
8		Hot Dog	$ 68.00	$ 74.00	$ 54.00	$ 196.00
9						
10		Totals	$ 216.25	$ 261.00	$ 218.25	$ 695.50
11						

A chart is nothing more than a visual representation of data, so first we need to enter in data. Let's assume you have kept track of all your hot dog stand sales from three concerts. Sales are also called revenues, and basically account for all the money taken in before subtracting all your costs (and thereby giving you your profit).

To create charts, we need to highlight information from our table. To make a column chart of the soda revenue from the three concerts, first highlight the cell range **B3:E4** and then click on the Insert tab. We are now ready to insert a chart. Click on the column chart command button and the chart below showing the **revenues** from selling soda at three concerts will magically appear.

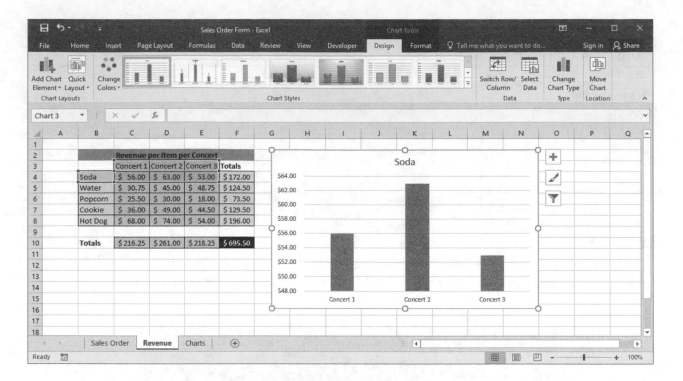

Notice how Excel automatically chooses titles for the columns and the chart from the highlighted information. When you click on a chart, you get two new tabs under *Chart Tools*: *Design* and *Format*. To chart all of the revenues we can highlight the entire range (not the totals) from B3:E8 and choose the basic 2-D clustered column chart. Right clicking on the bars allows us to change the fill from colors to patterned so they are differentiated in black and white printing.

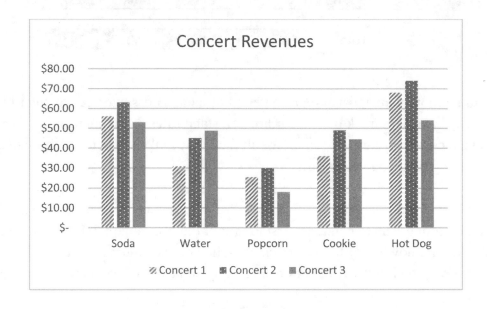

The same region of data can be highlighted and made into a stacked bar chart:

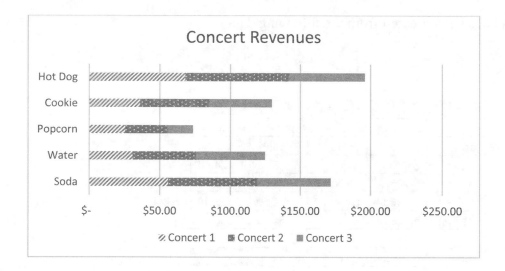

Pie charts are ubiquitous, but care must be used to ensure you do not include the total with all the parts of the pie. Including the total gives this embarrassing chart on the right:

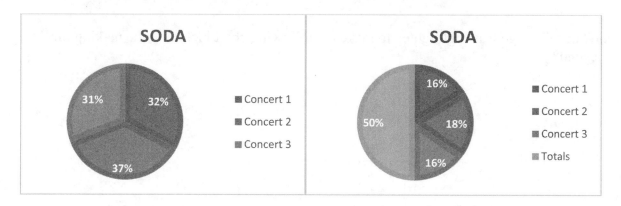

The charts have been created from the following data

	A	B	C	D	E	F
1						
2			Revenue per Item per Concert			
3			Concert 1	Concert 2	Concert 3	Totals
4		Soda	$ 56.00	$ 63.00	$ 53.00	$ 172.00
5		Water	$ 30.75	$ 45.00	$ 48.75	$ 124.50
6		Popcorn	$ 25.50	$ 30.00	$ 18.00	$ 73.50
7		Cookie	$ 36.00	$ 49.00	$ 44.50	$ 129.50
8		Hot Dog	$ 68.00	$ 74.00	$ 54.00	$ 196.00
9						
10		Totals	$ 216.25	$ 261.00	$ 218.25	$ 695.50
11						

1. What cell range was highlighted to create the second pie chart on the right above?

2. What cell range was highlighted to make the following chart including the making the legend?

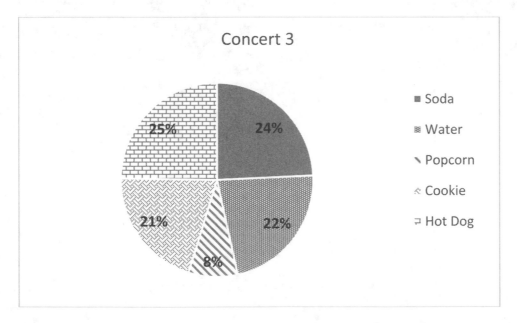

3. What does the stacked bar chart above tell that is not easy to read from the first column chart above?

4. What information do pie charts give that is not available in a column chart?

5. If you include the total with the parts in a pie chart why will the percentage for the total always be 50%?

Section 0.3 Reflections
Monitor Your Understanding

Can you describe the difference between a column chart and a pie chart?

Chapter 1 – An Excel-lent Approach to Relationships

Section 1.1 Functions

> **Objective 1** – Learn Function Basics
> **Objective 2** – Explore Financial Functions

> **Objective 1** – Learn Function Basics

Key Terms

Function	
Inputs	
Outputs	
Constant Function	
One-to-one (1-1) Function	
Domain	
Range	

Summary

In this course we will study relationships between quantities, with unique inputs determining one and only one output. Note that different inputs can all have the same output, we simply require each collection of inputs to determine a single output. Mathematicians refer to these relationships as functions so part of the goal is to familiarize you with the vocabulary associated with functions.

*As you read, jot down notes and questions. At the end of this section's Guided Worksheets there is a space for **Reflection** and **Monitoring Your Understanding**, where you can try and answer your questions after completing the guided activities and homework.*

Notes & Questions

Guided Practice Activity #1 – Reasoning About Functions

1. The annual cost of a car is a function of many inputs. List all of the inputs you can think of that determine the output of the annual car cost.

2. Determine which of the following are functions:

INPUTS	OUTPUT	FUNCTION?
States	Senators	
Senators	States	
States	Number of senators	
People	Anyone they been married to…	
People	Number of spouses	
US Citizens	Social Security Numbers	
SS#'s	US Citizens	
People	Birthdays	
Birthdays	People born on that day	
Students in this class	Shoe size	
Shoe Size	Students in this class with that shoe size	

3. Come up with a function between two quantities which remains a function when you switch the inputs and outputs…

Objective 2 – Explore Financial Functions

Key Terms

Principal	
Interest	
Period	
Balance	
Interest Rate	
Annual Percentage rate (APR)	
Periodic Rate	
Annual Percentage Yield (APY)	

Summary

Developing your financial literacy is an important part of this course and you will work on this in the examples and homework. Spreadsheets are an integral tool in dealing with financial matters and functions in general so you will also develop your spreadsheet fluency. Banks make money lending people money and charging them interest. A credit card is a type of loan that allows you to pay what you can afford each month (above a certain minimum). Paying just the minimum each month could lead to never paying off the loan, with the bank continuing to make money on the interest! A car loan is different in that the monthly payment and length of time to pay off are fixed and established when you take out the loan. Establishing good credit (your credit score) is crucial to a life of financial independence.

$$\text{Periodic Rate} = \frac{\text{APR}}{n}$$

	Assuming APR = 12%	
Period	n	APR/n
Quarterly	4	12%/4 = 3%
Monthly	12	12%/12 = 1%
Weekly	52	12%/52 = 0.231%
Daily	365	12%/365 = 0.033%

$$APY = \frac{\text{Total Interest in 1 Year}}{\text{Original Principal}}$$

As you read, jot down notes and questions. At the end of this section's Guided Worksheets there is a space for **Reflection** *and* **Monitoring Your Understanding**, *where you can try and answer your questions after completing the guided activities and homework.*

Notes & Questions

Guided Practice Activity #1 – Financial Literacy: Credit Card Loan

Let us assume you buy a Sony flat screen TV for your dorm room that costs $1,000. You make the purchase with a store credit card that has a 12% *Annual Percentage Rate* (APR). You do not have to make any monthly payments for the first year (sounds good!), but they will charge interest at the end of each month. How much do you owe at the end of the first year?

1. What is the Principal?

2. What is the monthly (periodic) interest rate?

3. Monthly interest is a function of what two inputs?

4. What is the interest charged for the first month?

> Double-click on the fill handle of the highlighted cells to quickly fill down to month 12 (you must have months 1-12 already filled in).

| B5 | | f_x | =D4 |

	A	B	C	D	E	F	
1	Buying a Flat Screen TV					Principal:	$ 1,000.00
2						APR:	12%
3	Month	Balance	Interest	New Balance		Period:	Month
4	1	$1,000.00	$ 10.00	$ 1,010.00		No. of periods	
5	2	$1,010.00	$ 10.10	$ 1,020.10		in 1 year:	12
6	3					Periodic Rate:	1.0%
7	4						

			Cell Reference Handbook
Relative	A4	Changes the row number when you fill up/down and changes the column letter when you fill left/right.	
Mixed	$A4	Changes the row number when you fill up/down and fixes the column letter when you fill left/right.	
Mixed	A$4	Fixes the row number when you fill up/down and changes the column letter when you fill left/right.	
Absolute	A4	Fixes the row number when you fill up/down and fixes the column letter when you fill left/right.	

5. In the spreadsheet above, what formula is in cell **C4**? **D4**?

6. Which formula involves an *absolute cell reference*?

7. Why do we need to fill in the second row before we fill the formulas down?

8. What do you owe at the end of the first year?

9. What is the APY?

Guided Practice Activity #2 – Periodic Rates

For each of the following situations determine the periodic rate (four decimal places of accuracy), and the amount of interest charged in the first period. Creating a spreadsheet to do this will greatly speed up your work!

1. Balance = $576 APR = 8% weekly

   a. Weekly rate?

   b. First week's interest?

2. Balance = $1396 APR = 3.5% daily

   a. Daily rate?

   b. First day's interest?

3. Balance = $18,500 APR = 14% monthly

   a. Monthly rate?

   b. First month's interest?

4. Balance = $18,500 APR = 1% monthly

   a. Monthly rate?

   b. First month's interest?

5. Balance = $450 APR = 240% monthly

   a. Monthly rate?

   b. First month's interest?

Excel Preview – 1.1 Heart Rate

▲	A	B	C	D	E	F	⦙
1							
2							
3		Average heart rate (in beats per minute) is given by the					
4		following formula:					
5		Heart Rate = (220 –Age) * 70%.					
6							
7		**a.)**					
8		Input Age:	18		a.) Enter a formula in the ⦙		
9		Output Heart Rate:	141.4		of *Age* (years). Be sure to u		
10					formula for an 18 year old ⦙		
11							

1. What formula would you type into cell **C9**?

2. According to this formula heart rate is a function of what input?

3. What are the units for the inputs and outputs for this function?

4. To make the following column chart how to we get the category x-axis labels to appear?

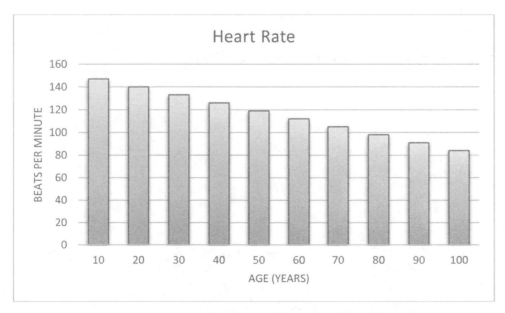

▲	A	B	C	D	E	F	G	H
1	Buying a Flat Screen TV						Principal:	$1,000.00
2		Part a:					APR:	12%
3	Month	Balance	Interest	Payment	New Balance		Period:	Month
4	1	$1,000.00	$ 10.00	$ 50.00			No. of periods	
5	2						in 1 year:	12
6	3						Periodic Rate:	1.0%
7	4							

1. Assuming we now make a $50 payment, what formula is entered into cell **E4**?

2. What will the balance be at the start of month 2?

3. What formula is entered in cell **B5**?

4. What formulas are entered in cells **C5**, **D5**, and **E5**?

5. Why do we need to fill in the formulas for month 2 before filling all of them down?

6. Can we fill down all the formulas for month 2 together? How?

Section 1.1 Reflections
Monitor Your Understanding

What was something you learned in this section that will make you think differently about something in your own life?

What was surprising about the TV Loan example to you?

How do banks make money lending people money?

When you take out a loan is it better to have a high APR or low APR? Why?

To pay off a credit card loan is it smart to make the smallest monthly payment allowed? Why?

Should you get a credit card and establish a solid credit history? Do you know what a "credit score is?

Objective 1 – Understand Order of Operations
Objective 2 – Evaluate Payment Function

Objective 1 – Understand Order of Operations

Key Terms	
Order of Operations	
PEMDAS	

Summary

Formulas in textbooks visually indicate fractions as numerators on top of denominators and exponents as superscripts. In Excel we have type in formulas without such visual cues, and thus need an agreed upon order in which to make computations to avoid ambiguity. Understanding the correct order of operations is crucial for your formulas in Excel to work properly.

CAUTION! When entering formulas into Excel you must use **Order of Operations**.

Examples:

       a. =3*5-2 = 13
       b. =3*(5-2) = 9
       c. =5+2*3 = 11 (not 21)

As you read, jot down notes and questions. At the end of this section's Guided Worksheets there is a space for **Reflection** *and* **Monitoring Your Understanding**, *where you can try and answer your questions after completing the guided activities and homework.*

Notes & Questions

Guided Practice Activity #1 – Order of Operations

1. For each of the following compute using proper order of operations without a calculator.

   a. 2–8/4+6*2

   b. 3^2+1

   c. 60/(2+8)-2^3*2

   d. 5+5/5+5

   e. 12/6*4

2. Write down how you would type these into Excel as a formula:

   a. $\dfrac{2+4}{3*6}-6$

   b. $5+\dfrac{2}{(8*4)^{3+1}}$

   c. $\left(\dfrac{9+2}{8-11}\right)^{4*6}-1$

Key Terms

Term	

Summary

In this objective we provide a real world example of a formula requiring careful use of order of operations when typing into Excel. Remember that car loans have fixed payments each month for a fixed length of time (the term). Loans of this type are called "amortized" loans, the root "amort" meaning death or killing off the loan.

As you read, jot down notes and questions. At the end of this section's Guided Worksheets there is a space for **Reflection** *and* **Monitoring Your Understanding**, *where you can try and answer your questions after completing the guided activities and homework.*

Notes & Questions

Guided Practice Activity #1 – Payment Formula Using Order of Operations

You have just bought a new VW Jetta for $17,254.38. You put down $2,254.38 of your savings so you only have to borrow $15,000. The auto dealership gets you a loan from a bank for 5 years at 6%, which you agree to and sign. What will be your fixed monthly payment?

The periodic payment is a function of 4 inputs:

INPUTS	OUTPUT
Principal (P)	Periodic payment (PMT)
APR	
Number of periods in 1 year (n)	
Number of years of the loan (t)	

$$PMT = \frac{P \times \dfrac{APR}{n}}{\left(1 - \left(1 + \dfrac{APR}{n}\right)^{-nt}\right)}$$

Use the following spreadsheet to answer the questions:

	A	B	C	D	E
1	**P**	**APR**	**n**	**t**	**PMT**
2	$ 15,000.00	6%	12	5	$ 289.99
3					

1. What is the term for this loan?

2. How many total payments will be made?

3. What will be the total amount paid?

4. What is the total amount interest paid?

5. What formula is entered in cell **E2**? Does APR/n need its own parentheses?

Excel Preview – 1.6 Taxes

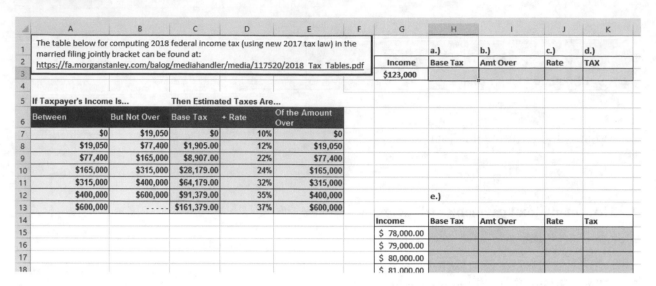

	A	B	C	D	E	F	G	H	I	J	K
								a.)	b.)	c.)	d.)
1	The table below for computing 2018 federal income tax (using new 2017 tax law) in the married filing jointly bracket can be found at:						Income	Base Tax	Amt Over	Rate	TAX
2	https://fa.morganstanley.com/balog/mediahandler/media/117520/2018_Tax_Tables.pdf										
3							$123,000				
4											
5	If Taxpayer's Income Is...		Then Estimated Taxes Are...								
6	Between	But Not Over	Base Tax	+ Rate	Of the Amount Over						
7	$0	$19,050	$0	10%	$0						
8	$19,050	$77,400	$1,905.00	12%	$19,050						
9	$77,400	$165,000	$8,907.00	22%	$77,400						
10	$165,000	$315,000	$28,179.00	24%	$165,000						
11	$315,000	$400,000	$64,179.00	32%	$315,000						
12	$400,000	$600,000	$91,379.00	35%	$400,000			e.)			
13	$600,000	-----	$161,379.00	37%	$600,000						
14							Income	Base Tax	Amt Over	Rate	Tax
15							$ 78,000.00				
16							$ 79,000.00				
17							$ 80,000.00				
18							$ 81,000.00				

1. For a married couple filing jointly who make $123,000 in 2018 what is their base tax?

2. How much more do they make than the "Of the Amount Over" value from column E for their row in the table?

3. Compute their estimated taxes by adding the base tax and the correct percentage of the amount they make over the cutoff value. Note there is no division here! The word "over" implies "more than".

4. Are you surprised by how much they pay in taxes?

5. To make the following scatterplot chart what columns have been highlighted?

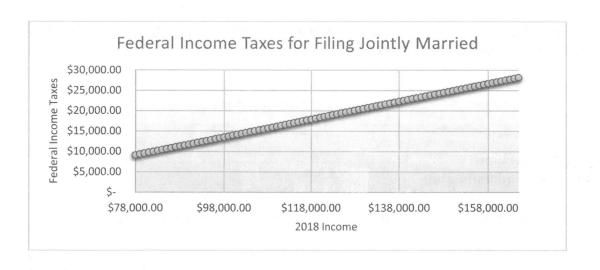

Excel Preview – 1.8 BMI

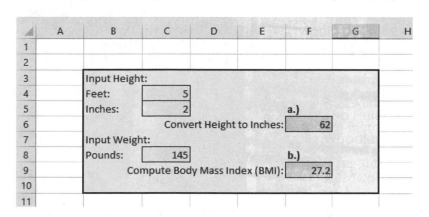

$$BMI = \frac{Weight\ (lbs)*0.45}{(Height\ (in.)*0.025)^2}$$

1. What formula is typed into **F6**?

2. Are parentheses needed in this formula?

3. What formula is typed into cell **F9**?

4. BMI is an index number and is meaningless without context. What else do we need to know in order to make sense out of the BMI computed in F9?

<u>**Section 1.2 Reflections**</u>
Monitor Your Understanding

What challenges are you facing in entering the complicated formulas in this section?

What did you learn about personal finance in this section that surprised you?

Section 1.3 Numerical and Graphical Representations: Tables and Charts

Objective 1 – Use Multiple Inputs to Understand What-If Analysis
Objective 2 – Convert Tables to Effective Graphs

Objective 1 – Use Multiple Inputs to Understand What-If Analysis

Key Terms

Cell reference	
Relative cell reference	
Absolute cell reference	
Mixed cell reference	

Summary

"What-if analysis" refers to using spreadsheets to model real world scenarios and asking "what-if" questions: What if the cost of fuel rises? What if I can get a better APR? What if I take out a 3 year loan instead of 6 years? What if I borrow $17,000 instead of $15,000? Using our spreadsheet model we can tweak the inputs and see what happens to the corresponding outputs. This typically involves functions with multiple inputs and we explore these now. Our cell references can become more complicated than simply relative and absolute. We can also have mixed cell references that change when we fill down but are fixed when we fill across (or vice versa).

Cell Reference Handbook		
Relative	A4	Changes the row number when you fill up/down and changes the column letter when you fill left/right.
Mixed	$A4	Changes the row number when you fill up/down and fixes the column letter when you fill left/right.
Mixed	A$4	Fixes the row number when you fill up/down and changes the column letter when you fill left/right.
Absolute	A4	Fixes the row number when you fill up/down and fixes the column letter when you fill left/right.

As you read, jot down notes and questions. At the end of this section's Guided Worksheets there is a space for **Reflection** *and* **Monitoring Your Understanding***, where you can try and answer your questions after completing the guided activities and homework.*

Notes & Questions

Guided Practice Activity – Monthly Car Payments as Function of Two Inputs

You have just bought a new VW Jetta for $17,254.38. You put down $2,254.38 of your savings so you only have to borrow $15,000. The auto dealership gets you a loan from a bank for 5 years at 6%, which you agree to and sign. What will be your fixed monthly payment?

The periodic payment is a function of 4 inputs:

INPUTS	OUTPUT
Principal (P)	Periodic payment (PMT)
APR	
Number of periods in 1 year (n)	
Number of years of the loan (t)	

$$PMT = \frac{P \times \dfrac{APR}{n}}{\left(1 - \left(1 + \dfrac{APR}{n}\right)^{-nt}\right)}$$

In the following spreadsheet we have already explored cell **E2** that is a function of the four inputs in cells **A2:D2**. Now we will consider the two-way table in **B6:H16**, where we are holding the principal, $P = \$15,000$, fixed along with the number of months in one year, $n = 12$. The monthly car payment becomes a function of two inputs: APR and term.

	A	B	C	D	E	F	G	H
1	P	APR	n	t	PMT			
2	$ 15,000.00	6%	12	5	$ 289.99			
3								
4			Table Showing Monthly Payments for Different APRs and Terms					
5			Term: Number of Years (t)					
6			1	2	3	4	5	6
7	APR	1%	$ 1,256.78	$ 631.53	$ 423.12	$ 318.92	$ 256.41	$ 214.73
8		2%	$ 1,263.58	$ 638.10	$ 429.64	$ 325.43	$ 262.92	$ 221.26
9		3%	$ 1,270.41	$ 644.72	$ 436.22	$ 332.01	$ 269.53	$ 227.91
10		4%	$ 1,277.25	$ 651.37	$ 442.86	$ 338.69	$ 276.25	$ 234.68
11		5%	$ 1,284.11	$ 658.07	$ 449.56	$ 345.44	$ 283.07	$ 241.57
12		6%	$ 1,291.00	$ 664.81	$ 456.33	$ 352.28	$ 289.99	$ 248.59
13		7%	$ 1,297.90	$ 671.59	$ 463.16	$ 359.19	$ 297.02	$ 255.74
14		8%	$ 1,304.83	$ 678.41	$ 470.05	$ 366.19	$ 304.15	$ 263.00
15		9%	$ 1,311.77	$ 685.27	$ 477.00	$ 373.28	$ 311.38	$ 270.38
16		10%	$ 1,318.74	$ 692.17	$ 484.01	$ 380.44	$ 318.71	$ 277.89
17								

1. What are the two specific inputs, APR and term, for the highlighted payment of $456.33?

2. What formula is entered in cell **C7** that will allow us to fill down and across? Note you may use the number 12 in your formula instead of the cell reference C2 since this is constant for monthly payments. In particular pay attention to what cell references are absolute, relative and mixed.

3. How much total interest is paid with the $289.99 monthly payment?

4. How much would you save in interest if you switched your 5 year 6% loan to 3 years at 6%?

5. What if you have a horrible credit score you might be forced to take out a 6 year loan at 10%. How much interest do you pay over the life of this loan?

6. What is fixed (column or row) by the following cell references?

F4	G$2	T1	$V85	J$21	$S99

Key Terms

Select Data	
Category x-axis labels	

Summary

A table of numbers can be visualized by highlighting a section of numbers and inserting a chart. This is an important point! A chart (or graph) is nothing more than a visual of a table of numbers. You must think carefully about which numbers should be visualized and how they will appear in the chart.

As you read, jot down notes and questions. At the end of this section's Guided Worksheets there is a space for **Reflection** *and* **Monitoring Your Understanding***, where you can try and answer your questions after completing the guided activities and homework.*

Notes & Questions

Guided Practice Activity – Effective Charts of Two Way Tables

To make an effective chart of a two way table we often simply highlight one column of data and then add appropriate labels. Let's assume we want to make a chart showing the monthly payments a 5 year loan of $15,000 as a function of the different APR's.

	A	B	C	D	E	F	G	H
1	P	APR	n	t	PMT			
2	$ 15,000.00	6%	12	5	$ 289.99			
3								
4			Table Showing Monthly Payments for Different APRs and Terms					
5			Term: Number of Years (t)					
6				1	2	3	4	5	6
7	APR	1%	$ 1,256.78	$ 631.53	$ 423.12	$ 318.92	$ 256.41	$ 214.73
8		2%	$ 1,263.58	$ 638.10	$ 429.64	$ 325.43	$ 262.92	$ 221.26
9		3%	$ 1,270.41	$ 644.72	$ 436.22	$ 332.01	$ 269.53	$ 227.91
10		4%	$ 1,277.25	$ 651.37	$ 442.86	$ 338.69	$ 276.25	$ 234.68
11		5%	$ 1,284.11	$ 658.07	$ 449.56	$ 345.44	$ 283.07	$ 241.57
12		6%	$ 1,291.00	$ 664.81	$ 456.33	$ 352.28	$ 289.99	$ 248.59
13		7%	$ 1,297.90	$ 671.59	$ 463.16	$ 359.19	$ 297.02	$ 255.74
14		8%	$ 1,304.83	$ 678.41	$ 470.05	$ 366.19	$ 304.15	$ 263.00
15		9%	$ 1,311.77	$ 685.27	$ 477.00	$ 373.28	$ 311.38	$ 270.38
16		10%	$ 1,318.74	$ 692.17	$ 484.01	$ 380.44	$ 318.71	$ 277.89
17								

1. What cell range should you highlight for this chart?

2. The following chart does not have correct x-axis labels. How do you get the APR's to show up on the x-axis?

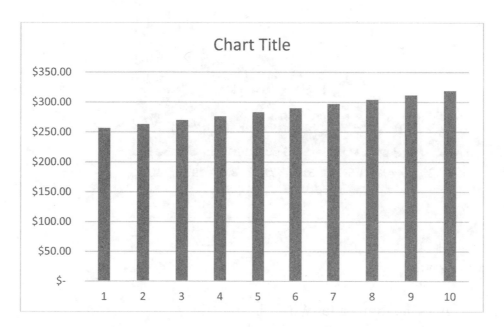

3. How do you add axis titles and the chart title?

4. How do you make the columns wider?

5. How do you make the vertical axis start at $250 to accentuate the difference in monthly payments?

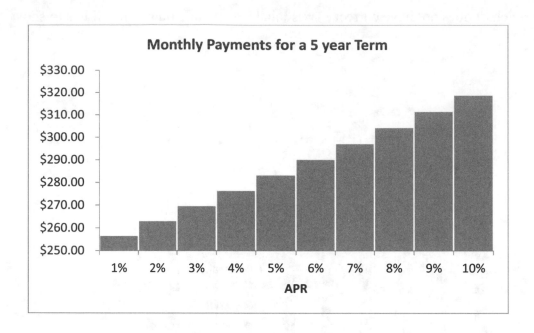

Monthly Payments for a 5 year Term

Excel Preview – 1.7 Wind Chill

The following screenshot gives the formula for Wind Chill.

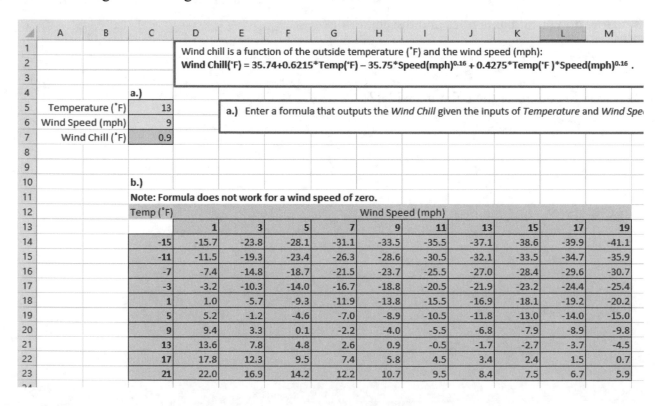

Wind chill is a function of the outside temperature (°F) and the wind speed (mph):

$$\text{Wind Chill}(°F) = 35.74 + 0.6215 \cdot \text{Temp}(°F) - 35.75 \cdot \text{Speed}(mph)^{0.16} + 0.4275 \cdot \text{Temp}(°F) \cdot \text{Speed}(mph)^{0.16}.$$

a.)

Temperature (°F)	13
Wind Speed (mph)	9
Wind Chill (°F)	0.9

a.) Enter a formula that outputs the *Wind Chill* given the inputs of *Temperature* and *Wind Spe*

b.)

Note: Formula does not work for a wind speed of zero.

Temp (°F)	Wind Speed (mph)									
	1	3	5	7	9	11	13	15	17	19
-15	-15.7	-23.8	-28.1	-31.1	-33.5	-35.5	-37.1	-38.6	-39.9	-41.1
-11	-11.5	-19.3	-23.4	-26.3	-28.6	-30.5	-32.1	-33.5	-34.7	-35.9
-7	-7.4	-14.8	-18.7	-21.5	-23.7	-25.5	-27.0	-28.4	-29.6	-30.7
-3	-3.2	-10.3	-14.0	-16.7	-18.8	-20.5	-21.9	-23.2	-24.4	-25.4
1	1.0	-5.7	-9.3	-11.9	-13.8	-15.5	-16.9	-18.1	-19.2	-20.2
5	5.2	-1.2	-4.6	-7.0	-8.9	-10.5	-11.8	-13.0	-14.0	-15.0
9	9.4	3.3	0.1	-2.2	-4.0	-5.5	-6.8	-7.9	-8.9	-9.8
13	13.6	7.8	4.8	2.6	0.9	-0.5	-1.7	-2.7	-3.7	-4.5
17	17.8	12.3	9.5	7.4	5.8	4.5	3.4	2.4	1.5	0.7
21	22.0	16.9	14.2	12.2	10.7	9.5	8.4	7.5	6.7	5.9

1. In cell **D14** what are the inputs for the output of -15.7 degrees?

2. What formula is in cell **D14** that you can fill down and across? Pay attention to what should be a relative versus absolute or mixed cell reference.

   Wind Chill(°F) =
   35.74+0.6215*Temp(°F) − 35.75*Speed(mph)$^{0.16}$ + 0.4275*Temp(°F)*Speed(mph)$^{0.16}$

3. What cell range was highlighted to make the following line chart?

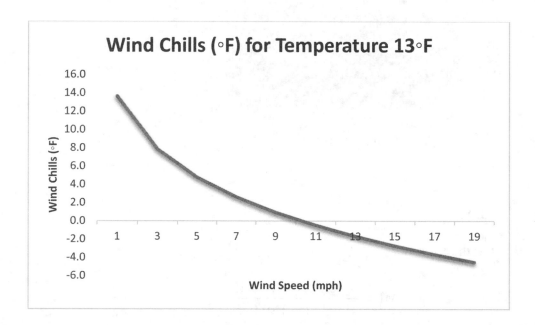

4. What cell range was highlighted to make the following surface chart?

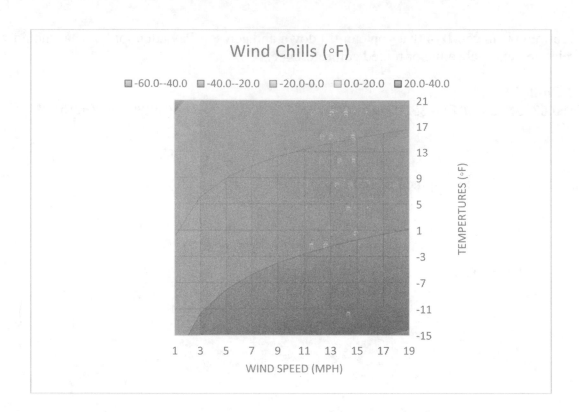

Excel Preview – 1.8 BMI

BMI is calculated using the formula:

$$BMI = \frac{Weight\ (lbs) * 0.45}{(Height\ (in.) * 0.025)^2}$$

			c.)									
20												
21				C	D	E	F	G	H	I	J	K	L
22		Height (in)						Weight (lbs)					
23				105	110	115	120	125	130	135	140	145	150
24			60	21.0	22.0	23.0	24.0	25.0	26.0	27.0	28.0	29.0	30.0
25			61	20.3	21.3	22.3	23.2	24.2	25.2	26.1	27.1	28.1	29.0
26			62	19.7	20.6	21.5	22.5	23.4	24.3	25.3	26.2	27.2	28.1
27			63	19.0	20.0	20.9	21.8	22.7	23.6	24.5	25.4	26.3	27.2
28			64	18.5	19.3	20.2	21.1	22.0	22.9	23.7	24.6	25.5	26.4
29			65	17.9	18.7	19.6	20.4	21.3	22.2	23.0	23.9	24.7	25.6
30			66	17.4	18.2	19.0	19.8	20.7	21.5	22.3	23.1	24.0	24.8
31			67	16.8	17.6	18.4	19.2	20.0	20.9	21.7	22.5	23.3	24.1
32			68	16.3	17.1	17.9	18.7	19.5	20.2	21.0	21.8	22.6	23.4
33			69	15.9	16.6	17.4	18.1	18.9	19.7	20.4	21.2	21.9	22.7
34													

1. What formula would you type into cell **C24** that can be filled down and across?

2. What cell range would be highlighted to make the following chart?

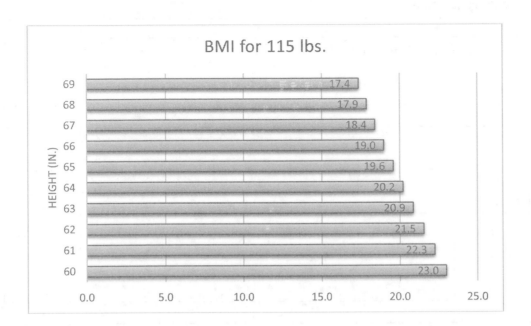

BMI for 115 lbs.

Section 1.3 Reflections
Monitor Your Understanding

What is one important tip you learned for making effective charts?

Section 1.4 Spotlight on Statistics

| **Objective 1** – Learn Descriptive Statistics |
| **Objective 2** – Create Histograms |

| **Objective 1** – Learn Descriptive Statistics |

Key Terms

Mean	
Median	
Mode	
Standard deviation	
Max	
Min	
Range	
Count	

Summary

The discipline of statistics deals with collecting, organizing, analyzing, and displaying data. A **statistic** is a number you compute related to a data set, which gives you information about that data set. Numeric data such as ages is called quantitative, while non-numeric data like gender is qualitative. Quantitative data has three essences when ordered from smallest to largest:

1. **Shape**: a column chart (histogram) showing how many values lie in given intervals or bins.
2. **Center**: the middle of the data as measured by the mean, median, and mode.
3. **Spread**: the variability of the data as measured by the maximum and minimum values and the standard deviation, or average distance of the values from the mean.

As you read, jot down notes and questions. At the end of this section's Guided Worksheets there is a space for **Reflection** *and* **Monitoring Your Understanding***, where you can try and answer your questions after completing the guided activities and homework.*

Notes & Questions

Guided Practice Activity – Age Distribution

The table and chart give the distribution of ages in the U.S. in 2016.

Bins	Males	Females
0 -- 9	20,616,776	19,740,060
10 -- 19	21,320,578	20,427,654
20 -- 29	23,122,539	22,149,373
30 -- 39	21,344,706	21,215,558
40 -- 49	20,152,276	20,491,598
50 -- 59	21,413,590	22,405,574
60 -- 69	17,252,551	19,050,568
70 -- 79	9,177,809	11,000,333
80 -- 89	3,916,541	5,834,320
90 -- 99	745,215	1,667,998
100 +	16,342	65,554

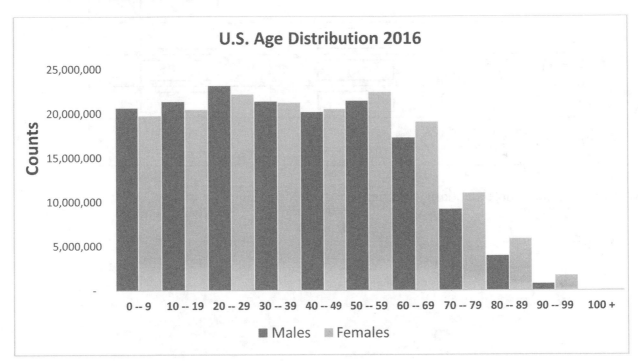

1. What are the heights of the columns for the first bin in which women outnumber men? What do these heights represent?

2. The text indicates the median age is about 38 years old. What does this mean?

3. Compute the number of people from 0 – 39 and 40 and above. Are these close to being equal?

<u>**Guided Practice Activity**</u> – Latte

You create a data **sample** of the costs of a latte from visiting four cafés: {$1, $3, $5, $7}.

1. Evaluate the **mean**.

2. Fill in the following table.

	Values	**Deviations**	**Deviations²**
x_i	x_i	$x_i - \overline{X}$	$\left(x_i - \overline{X}\right)^2$
x_1	$1		
x_2	$3		
x_3	$5		
x_4	$7		
Sums:	$\sum x_i = \Box$	$\sum\left(x_i - \overline{X}\right) = \Box$	$\sum\left(x_i - \overline{X}\right)^2 = \Box$

3. Compute the sample standard deviation using the formula:

$$s = \sqrt{\frac{\sum\left(x_i - \overline{X}\right)^2}{n-1}}$$

Key Terms

Histogram	
Bin	
Frequency	
Distribution	
Statistics	

Summary

The Age distribution chart is an example of a histogram, showing how many data values fall into given bins or intervals. This distribution gives the shape of your data.

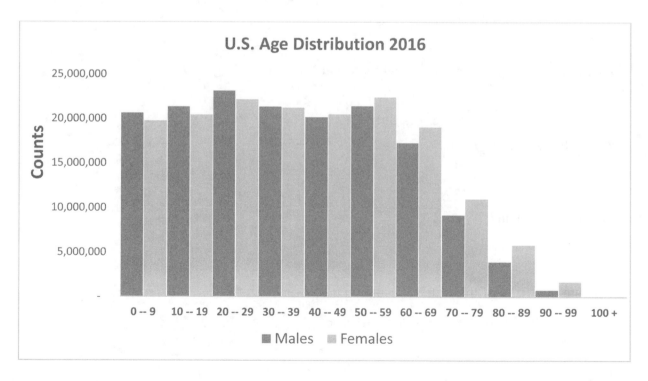

As you read, jot down notes and questions. At the end of this section's Guided Worksheets there is a space for **Reflection** *and* **Monitoring Your Understanding**, *where you can try and answer your questions after completing the guided activities and homework.*

Notes & Questions

Guided Practice Activity – Coffee

You sit outside the campus café and survey the first 31 people who leave, asking how many cups of coffee they drink in a day.

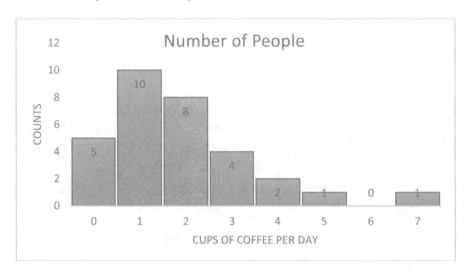

1. What does the height of each column represent?

2. What is the mode?

3. What is the median?

4. Compute the mean of this data set using your answer from #1.

5. Look at the following table that is shown in the video. Try to make sense of where the numbers are coming from.

Values	Frequency	A*B	deviations	dev^2	B*E
0	5	0	-1.87	3.50	17.50
1	10	10	-0.87	0.76	7.59
2	8	16	0.13	0.02	0.13
3	4	12	1.13	1.27	5.10
4	2	8	2.13	4.53	9.07
5	1	5	3.13	9.79	9.79
6	0	0	4.13	17.05	0.00
7	1	7	5.13	26.31	26.31
	Mean:	1.87			75.48
Mean:	1.87				2.52
				Std Dev:	1.59

Guided Practice Activity – CAT Scores

We will use the *CAT Scores* worksheet in *Data Sets*.

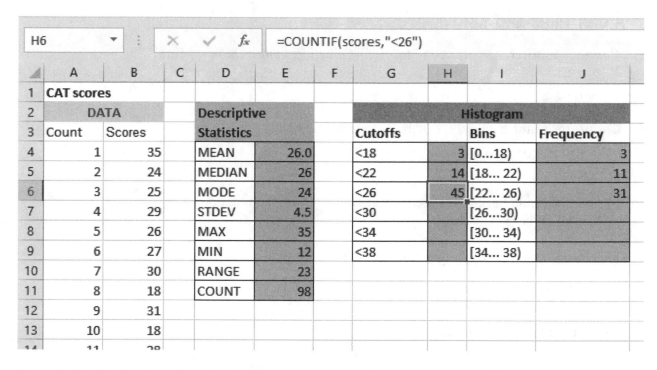

Descriptive Statistics: Note that the name of each function in Excel is as given in the spreadsheet, except for the mean which uses the **AVERAGE** function, and the range which is **MAX - MIN** . Enter the appropriate functions into the spreadsheet to compute the following, and write the formula you would type into Excel after each definition (note there are 98 scores listed in column **B** and the cell range **B4:B101** has been named *scores*).

1. *Mean:* the arithmetic average.

2. *Median:* the middle of the data set (half above, half below, 50th percentile).

3. *Mode:* the most frequently occurring value.

4. *Standard Deviation:* the "average distance" of the data values from the mean.

5. *Max:* the largest value.

6. *Min:* the smallest value.

7. *Range:* the difference between largest and smallest values (Max – Min).

8. *Count:* the number of values (usually referred to as N).

   Note the formula bar in the spreadsheet above giving the function
   =COUNTIF(scores,"<26").

9. What is the output 45 telling us in cell **H6**?

10. How do you *name* a cell range *scores*?

11. The **COUNTIF** function has two arguments, what are they called in general?

12. What formula should be typed into cell **H7**?

13. Why is there 14 in cell **H5** but only 11 in cell **J5**?

14. What formula is in cell **J6**?

Finish entering formulas in the Histogram box and create a column chart of the last two columns, *Bins* and *Frequency*. This is called a Histogram.

Complete the histogram below by drawing in the columns for the last three bins.

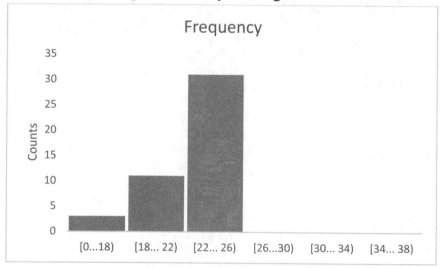

	A	B	C	D	E	F	G	H	I	J	K	L	M	N
1	Table 1.3—SUMMARY OF RECEIPTS, OUTLAYS, AND SURPLUSES OR DEFICITS (–) IN CURRENT DOLLARS, CONSTANT (FY 2009) DOLLARS, AND AS PERCENTAGES OF GDP: 1940–2021													a.)
2					(dollar amounts in billions)								Mean:	17.0
3		In Current Dollars			In Constant (FY 2009) Dollars			Addendum: Composite Deflator	As Percentages of GDP				b.)	c.)
4	Fiscal Year	Receipts	Outlays	Surplus or Deficit (–)	Receipts	Outlays	Surplus or Deficit (–)		Receipts	Outlays	Surplus or Deficit (–)		Deviation	Deviation^2
5	1940	6.5	9.5	-2.9	94.0	135.8	-41.9	0.0697	6.7	9.6	-3.0		=I5-AVG	106.54
6	1941	8.7	13.7	-4.9	113.3	177.5	-64.3	0.0769	7.5	11.7	-4.3		-9.5	90.66
7	1942	14.6	35.1	-20.5	167.8	402.9	-235.1	0.0872	9.9	23.8	-13.9		-7.1	50.72
8	1943	24.0	78.6	-54.6	249.0	814.9	-565.9	0.0964	13.0	42.6	-29.6		-4.0	16.17
9	1944	43.7	91.3	-47.6	493.2	1,029.4	-536.2	0.0887	20.5	42.7	-22.2		3.5	12.10
10	1945	45.2	92.7	-47.6	541.5	1,111.7	-570.2	0.0834	19.9	41.0	-21.0		2.9	8.28
11	1946	39.3	55.2	-15.9	466.7	656.0	-189.3	0.0842	17.2	24.2	-7.0		0.2	0.03
12	1947	38.5	34.5	4.0	402.9	360.8	42.0	0.0955	16.1	14.4	1.7		0.9	0.85

Yikes! Lots of information here. It is crucial you understand what your data means **before** trying to answer questions about it. If you do not understand what a *Receipt* is or what *Surplus* means it will be very difficult for you to answer questions about *Receipts* and *Surpluses*.

1. What is a Receipt for the Federal Government?

2. What does the number 6.5 in cell B5 mean?

3. What does the 9.5 in cell B6 mean?

4. What does the -2.9 in cell D6 mean?

5. What is the difference between Current Dollars and Constant (FY 2009) Dollars?

6. What does GDP stand for?

7. Ok, you are now ready to compute the standard deviation of column I using the formula. But first you need to name the mean in cell N2? How do you name a cell AVG?

Excel Preview – 1.12 Lobsters

	A	B	C	D	E	F	G	H	I
1	Historical Maine Lobster Landings								**a.)**
2	YEAR	SPECIES	METRIC	TONS	POUNDS	POUNDS(millions)	VALUE	VALUE(millions)	PRICE/LB
3	1950	LOBSTER,	AMERICAN	8324.6	18,352,600	18.3526	$6,412,311	6.4123	$0.35
4	1951	LOBSTER,	AMERICAN	9416.4	20,759,500	20.7595	$7,214,107	7.2141	$0.35
5	1952	LOBSTER,	AMERICAN	9088.3	20,036,300	20.0363	$8,511,821	8.5118	$0.42
6	1953	LOBSTER,	AMERICAN	10115.3	22,300,300	22.3003	$8,411,229	8.4112	$0.38
7	1954	LOBSTER,	AMERICAN	9828.3	21,667,700	21.6677	$8,087,156	8.0872	$0.37

1. To compute the price per pound what formula is entered in cell **I3**?

2. What does the number 47 mean in the following table that is linked to the POUNDS data **E3:E61**?

Cutoffs	Counts	Bins (in millions)	Frequency
20,000,000	17	0..20	17
30,000,000	43	20..30	26
40,000,000	47	30..40	4
50,000,000	50	40..50	3
60,000,000	53	50..60	3
70,000,000	57	60..70	4
80,000,000	59	70..80	2

3. What function is typed in the cell **L25** to give 47?

4. What does the number 4 (in same row as 47) represent?

5. What function is typed in the cell **N25** to give 4?

6. What data is highlighted to make the following histogram?

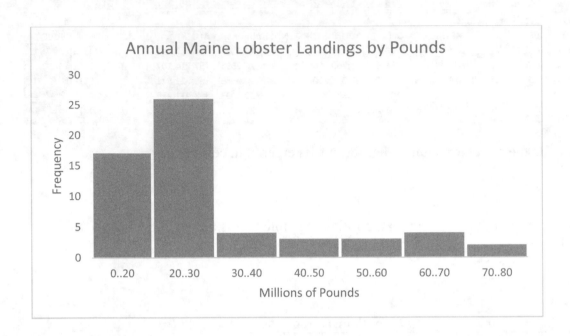

Section 1.4 Reflections
Monitor Your Understanding

1. Was it difficult for you to work with data sets like the Receipts or Lobster data? Why is this challenging?

2. What is meant by the "shape" of your data?

Chapter 2 – Ratios and Proportions

Section 2.1 Ratios and Proportions

Objective 1 – Understand Ratios
Objective 2 – Use Ratio Tables to Set Up Proportions
Objective 3 – Learn Ratio Vocabulary
Objective 4 – Contrast Part-to-Part and Part-to-Whole Ratios

Objective 1 – Understand Ratios

Key Terms

Ratio	

Summary

In Chapter 1 on functions we said that we will study relationships between quantities. Ratios are the most basic form of a relationship, simply comparing the relative size of two quantities. The number of births in 2017 in the U.S. was 3,853,472[1]. Is this a lot of births? To make sense of this statistic we can make comparisons across time, for example, this number is down 2% from 2016 and the lowest number in 30 years; or across geography, comparing it to the number of births in another country. Mexico only had 2,320,430 births[2] in 2017, but their population was 129,163,276 people compared to the U.S. population of 324,459,463 on July 1, 2017[3]. We could compute the number of births per 1,000 people, 11.88 for the U.S. and 17.97 for Mexico, telling us Mexico has more babies per people crawling around. In this sense, Mexico has "more" births, even though our absolute number of births (3,853,472) is larger. Only women of child bearing age can have babies of course, so maybe we should be comparing births to the number of these women instead of the entire population! As you can see, there is more to comparing numbers than meets the eye. This chapter will lay the foundation for how we go about making these comparisons, namely by using ratios.

---

[1] https://www.cdc.gov/nchs/data/vsrr/report004.pdf
[2] http://worldpopulationreview.com/countries/births-by-country/
[3] https://en.wikipedia.org/wiki/List_of_countries_by_population_(United_Nations)

As you read, jot down notes and questions. At the end of this section's Guided Worksheets there is a space for **Reflection** *and* **Monitoring Your Understanding***, where you can try and answer your questions after completing the guided activities and homework.*

Notes & Questions

Guided Practice Activity #1 – Thing 1 and Thing 2

Ratios compare the relative size of two quantities that I refer to as Thing 1 and Thing 2. Identifying Thing 1 and Thing 2 is trickier than it seems, often because ratios are presented as single numbers, like the birth rate of the U.S. in 2017 was 11.88 hides that there are two quantities being compared: the number of births (Thing 1) and people (Thing 2).

1. What are the two quantities being compared in the following ratios? Be careful not to confuse units with the actual quantity. The gas mileage of a Subaru Outback, 35 miles per gallon, is a ratio but the quantities are not *miles* and *gallons*. Those are the **units** for the underlying **quantities** of *distance* (Thing 1) and amount of *gas* (Thing 2).

Ratio	Thing 1		Thing 2
GPA is 2.74		:	
Only 1 U.S. Adult in 5 could calculate interest		:	
13% of U.S. Adults are numerate		:	
APR = 6%		:	
Speed is 75 mph		:	
Half of 17 year olds don't know enough math to work in auto plant		:	
Math SAT scores for Elem. Ed. majors compared to Natl. Avg. is 483 vs. 515		:	
You borrow $300 and pay $350 at the end of the month		:	
52% of freshman say their emotional health is above average		:	
15.5% of students who started repaying student loans in 2012 were in default by 2016		:	
50.3% of Americans projected to be minorities in 2044		:	
Apple Inc. reported $9.21 earnings per share in 2017		:	

Key Terms

Proportion	
Ratio table	

Summary

In the first objective we introduced the birth rate for the U.S. of 11.88 births per 1,000 people, and mentioned we could also compute a fertility rate comparing births to the number of women aged 15 – 44. The choice of per 1,000 was arbitrary, we just as easily could report births per 1 person or per 1 million people. Keeping track of all the numbers we are comparing gets confusing, and we use ratio tables to organize the information. To compute the new ratio we use a fundamental technique:

1. Set up a proportion.
2. Cross multiply and solve for the unknown.

As you read, jot down notes and questions. At the end of this section's Guided Worksheets there is a space for **Reflection** *and* **Monitoring Your Understanding***, where you can try and answer your questions after completing the guided activities and homework.*

Notes & Questions

Guided Practice Activity #1 – Ratio Tables and Proportions.

The number of births in 2017 in the U.S. was 3,853,472[4]. Is this a lot of births? To make sense of this statistic we can make comparisons to the U.S. population of 324,459,463 on July 1, 2017[5] and compute the birth rate per 1,000 people. Given the fertility rate of 60.2 births per 1,000 women aged 15 – 44 in 2017 we can work backwards and compute the number of child bearing women. We organize all of this information in a ratio table.

Births	Population	Women (15 – 44 years of age)
3,853,472	324,459,463	*w*
b	1,000	
60.2		1,000

---

[4] https://www.cdc.gov/nchs/data/vsrr/report004.pdf
[5] https://en.wikipedia.org/wiki/List_of_countries_by_population_(United_Nations)

We make comparisons across the rows:

$$3{,}853{,}472 \text{ births} : 324{,}459{,}463 \text{ people} : w \text{ women } (15-44).$$

Let's start with easier numbers to get a feel for the solution technique, then come back and solve for the unknown quantities.

1. Compute the ratio of births to 1,000 people and the number of women in this population.

Births	Population	Women (15 – 44 years of age)
4	500	
	1,000	
12		900

2. Compute the number of women (15 – 44) in this population

3. Ok, we are now ready to return to the original ratio table. Solve for *w* (whole number) and *b* (2 decimal places)

Births	Population	Women (15 – 44 years of age)
3,853,472	324,459,463	*w*
b	1,000	
60.2		1,000

Let's now complete the table from the Chapter.

Tuition and Fees 2013-14

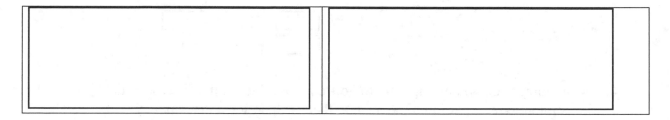

	Private 4-year	Public 4-year (In-State)	Public 2-year
	$30,094	$8,893	$3,264
Multiply by … ?		1	
	100		
			1

We have scaled this quantity to 1.

4. Determine the ratio of Private 4-year to Public 4-year In-State with the second quantity scaled to 1 using the fundamental solution technique to compute ratios:

   1. Set up proportion:

   2. Cross multiply and solve!

5. Fill in the values of all the boxes in the ratio table above. For the greatest accuracy you should always try to use the original statistics in the 1st row of numbers in the table.

Tuition and Fees 2013-14

Private 4-year	Public 4-year (In-State)	Public 2-year
$30,094	$8,893	$3,264
☐	1	☐
100	☐	☐
☐	☐	1

6. What number do you multiply by to go from the values in the 2nd row to the 3rd row? From 3rd column to the 1st column? Note that there will be some discrepancy with the values obtained by doing this and those in the table due to rounding.

7. Fill in the numbers in the bar chart below.

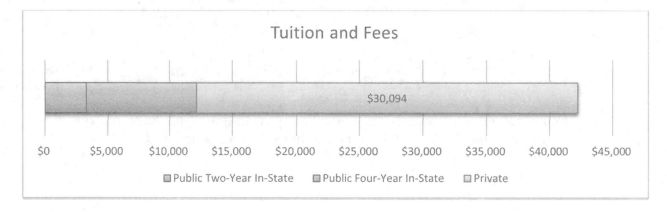

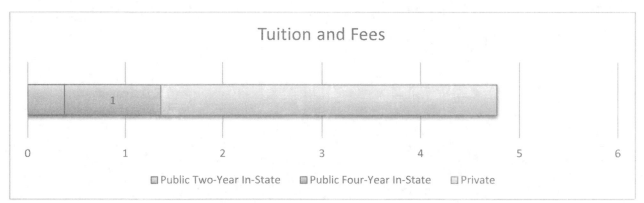

Guided Practice Activity #2 – Taming the River

Ratios seem so harmless when presented in math classes, 2 fish is to 3 bunnies as 8 fish is to 12 bunnies. In the real world we run into difficult tables of information like the one below. Test your quantitative literacy skills and see if you can answer the following questions. Good luck!

Table 2.1 Taming the River
Number of courses taken during freshman and sophomore year
(standardized cohort of 1,000 students)

Division and Department	Whites	Asians	Latinos	Blacks
Humanities				
Art and Art History	541	459	505	291
Classics	84	52	42	51
Comparitive Literature	55	57	102	99
English	1,567	1,414	1,536	1,589
Foreign Language	1,685	1,517	1,805	1,714
Spanish	593	337	918	807
Other	1,092	1,180	887	990
History	827	584	811	751
Music	782	823	655	657
Philosophy	519	426	446	406
Religion	357	249	288	248
Rhetoric	22	18	16	21
Theater and Media Arts	298	160	306	278
Other Humanities	768	357	893	801
Total	7,505	6,116	7,405	6,906
Ratio to Whites	1.00	0.81	0.99	0.92

1. What does the number 541 mean in the first column?

2. Compute the ratio of English to Art classes for each column, and scale the ratios so that the second quantity is 1. Round to 2 decimal places.

3. Compute the ratio of Art to English for each column, and scale the ratios so that the second quantity is 100. Round to 1 decimal place.

4. How was the Ratio to Whites computed? What are the two quantities involved?

5. Create three ratios of your choosing and scale the second quantity to a nice number ☺

Summary

Quantitative literacy implies an ability to communicate effectively about the ratios you are encountering and computing. There are several phrases that indicate ratios for the following data:

Private Tuition and Fees	Public Four-Year Tuition and Fees In-State	Public Two-Year Tuition and Fees In-State
$30,094	$8,893	$3,264
3.4	1	0.37
100	29.6	10.8

1. *For every* $100 a full-time undergraduate student spent on tuition and fees at a private four-year school in 2013-14, a public four-year in-state student spent $29.60.
2. Full-time undergraduate students at private four-year schools spent 3.4 *times as much* as students at public four-year in-state schools on tuition and fees in 2013-14.

3. Tuition and fees at public two-year schools were $\frac{1}{9}$ *of* the tuition and fees at private four-year schools in 2013-14.

As you read, jot down notes and questions. At the end of this section's Guided Worksheets there is a space for **Reflection** *and* **Monitoring Your Understanding**, *where you can try and answer your questions after completing the guided activities and homework.*

Notes & Questions

<u>**Guided Practice Activity #1**</u> – Ratio Literacy

Use the following data for tuition and fees for 2017–18 to create sentences as instructed.

Private Tuition and Fees	Public Four-Year Tuition and Fees In-State	Public Two-Year Tuition and Fees In-State
$34,740	$9,970	$3,570

1. Write a sentence that tells how many *times as much* students at public four-year in-state schools spent on tuition and fees in 2017-18 than public two-year in-state students.

2. Write a sentence that communicates public two-year tuition and fees in-state as *a fraction of* private tuition and fees in 2017-18.

3. Write a sentence that communicates how many dollars public four-year in-state students spend on tuition and fees *for every* $100 a private four-year student spends.

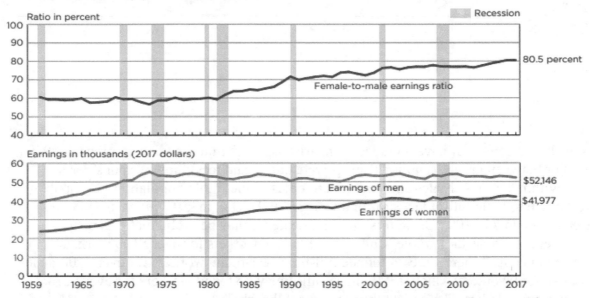

Figure 2.
Female-to-Male Earnings Ratio and Median Earnings of Full-Time, Year-Round Workers
15 Years and Older by Sex: 1960 to 2017

Note: The data for 2013 and beyond reflect the implementation of the redesigned income questions. The data points are placed at the midpoints of the respective years. Data on earnings of full-time, year-round workers are not readily available before 1960. For information on recessions, see Appendix A. For information on confidentiality protection, sampling error, nonsampling error, and definitions, see <www2.census.gov/programs-surveys/cps/techdocs/cpsmar18.pdf>.
Source: U.S. Census Bureau, Current Population Survey, 1961 to 2018 Annual Social and Economic Supplements.

4. How many times as much was male median earnings than women's median earnings in 2017?

5. For every $100 a male earned in median earnings, how much did a women earn in 2017?

6. Why is it important that the earnings are in 2017 dollars?

7. How does this graphic show "middle-class incomes are treading water"?

Key Terms

Part-to-part ratio	
Part-to-whole ratio	

Summary

The ratios we have seen so far have included comparisons across time (median earnings in 1970 versus 2017) or between groups (public versus private tuition and fees); now we consider comparisons within a group. Males and females are **parts** of the whole population. Computing the ratio of female to male full-time year round workers in the U.S. in 2017 gives us 49.3 : 66.4, which we can scale to 74 : 100. Using this ratio in a sentence may be problematic as people want to interpret the 100 as the "whole" and say "74 female workers for every 100 workers". The correct sentence is: "There were 74 female full-time year-round workers for every 100 male full-time year-round workers, which is 74 female full-time year-round workers for every 174 full-time year-round workers." You add the parts to get the whole.

Figure 3.
Total and Full-Time, Year-Round Workers With Earnings by Sex: 1967 to 2017

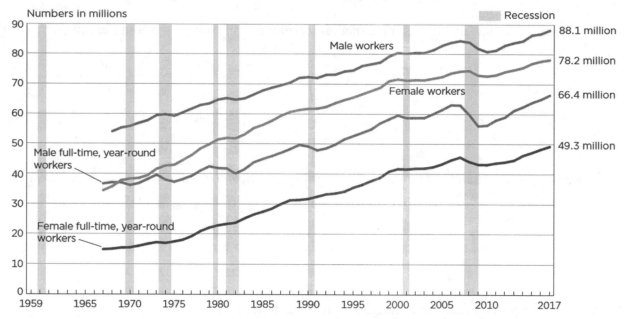

Note: The data for 2013 and beyond reflect the implementation of the redesigned income questions. The data points are placed at the midpoints of the respective years. Data on number of workers are not readily available before 1967. Data are for people aged 15 and older beginning in 1980 and people aged 14 and older for previous years. Before 1989, data are for civilian workers only. For information on recessions, see Appendix A. For information on confidentiality protection, sampling error, nonsampling error, and definitions, see <www2.census.gov/programs-surveys/cps/techdocs/cpsmar18.pdf>.

Source: U.S. Census Bureau, Current Population Survey, 1968 to 2018 Annual Social and Economic Supplements.

As you read, jot down notes and questions. At the end of this section's Guided Worksheets there is a space for **Reflection** *and* **Monitoring Your Understanding***, where you can try and answer your questions after completing the guided activities and homework.*

Notes & Questions

Guided Practice Activity #1 – Student Expenses

Use the following data for public two-year in-district student expenses for 2017-18 to answer the questions.

Tuition and Fees	Room and Board	Books and Supplies	Transportation	Other Expenses	Total
$3,570	$8,400	$1,420	$1,780	$2,410	$17,580
a	☐	*b*	☐	☐	100
c	☐	1	☐	☐	☐

1. Compute the part-to-whole ratio of tuition and fees to total expenses with the second quantity scaled to 100. Use the ratio in a sentence.

2. Compute the part-to-part ratio of tuition and fees to books and supplies. Use the ratio in a sentence.

3. Compute the part-to-part ratio of tuition and fees to books and supplies with the second quantity scaled to 1. Use the ratio in a sentence.

4. Complete the rest of the ratio table above.

Excel Preview – 2.5 Deaths

	A	B	C	D	E	F
1		2010 Population (US Census 2010):	308,745,538	a.)	b.)	c.)
2	Rank	Cause of Death	Number	Proportion of Total Deaths	Percent of Total Deaths	Crude Death Rate (2010)
3		All causes	2,468,435	Decimal	100.0	799.5
4	1	Diseases of heart	597,689			
5	2	Malignant neoplasms	574,743			
6	3	Chronic lower respiratory diseases	138,080			
7	4	Cerebrovascular diseases	129,476			
8	5	Accidents (unintentional injuries)	120,859			

1. Compute the part-to-whole ratio of heart disease deaths to all causes, with second quantity scaled to 1. What formula would you type into cell **D4** that can be filled down?

2. Compute the part-to-whole ratio of heart disease deaths to all causes, with second quantity scaled to 100. What formula would you type into cell **E4** that can be filled down?

3. Compute the part-to-whole ratio of heart disease deaths to the population, with second quantity scaled to 100,000. What formula would you type into cell **F4** that can be filled down?

	G	H	I	J	K	L	M	N	O
14									
15						d.)			
16					Age Adjusted Death Rate				
17			Percent Change	Ratio					
18			2010	2009-2010	Male	Female	Total	Male	Female	Black to White
19	Rank		747.0	−0.3	1.4	1.0	2.4			1.2
20	1		179.1	−2.0	1.6	1.0				1.3
21	2		172.8	−0.4	1.4	1.0				1.2
22	3		42.2	−1.2	1.3	1.0				0.7
23	4		39.1	−1.3	1.0	1.0				1.4

4. For each cause of death we are given the part-to-part ratio of male-to-female deaths. We can use the actual number of deaths from above to compute the number of male and female deaths for each cause.

CAUTION! Note that the government just reports the male to female ratio as single number, like the overall ratio of 1.4 in cell **J19**. We infer the second quantity is 1 and add the parts to get the total of 2.4 in cell **L19**.

Compute the total number of male deaths for all causes using this ratio.

5. What formula can you enter in cell **M19** that can be filled down (assuming the totals in column L are filled in)?

6. To compute the female deaths in cell N19 what are two possible formulas you can enter and fill down?

Excel Preview – 2.9 MF Workforce

	A	B	C	D	E	F	G	H	I
1	United States					a.)	b.)	b.)	c.)
2	Year	# Workers	Male	:	Female		Number of Workers (millions)		
3		(millions)	Employment Ratio			Whole	Male	Female	Ratio
4	1870	12.5	85	:	15				
5	1880	17.4	85	:	15				
6	1890	23.3	83	:	17				
7	1900	29.1	82	:	18				
8	1910	38.2	79	:	21				
9	1920	41.6	79	:	21				

1. We are given the part-to-part ratio of male to female workers. To compute the whole what formula do we enter in cell **F4**? What is the whole?

2. How many male workers were there in 1870? What formula is entered in cell **G4** that can be filled down?

3. How many female workers were there in 1870? What formula is entered in cell **H4** that can be filled down?

4. Compute the female to male ratio of workers and scale the second quantity to 100. What formula can be entered in cell **I4** that can be filled down?

5. Can the female to male ratios in column I ever be greater than 100? Why or why not?

Section 2.1 Reflections
Monitor Your Understanding

What was something you learned in this section that will make you think differently about something in your own life?

What was surprising about any of the statistics in this section?

Section 2.2 Weighted Averages

Objective 1 – Compute Weighted Averages

Key Terms	
Weighted average	

Summary

The 2015-16 tuition, fees, room and board in the table below are said to be "enrollment weighted" as indicated by the note from the National Center of Education Statistics:

Private Four-Year	Public Four-Year	Public Two-Year
$39,529	$19,189	$9,939

NOTE: Data are for the entire academic year and are average charges for full-time students. Tuition and fees were **weighted by the number of full-time-equivalent undergraduates**, but were not adjusted to reflect student residency. **SOURCE:** U.S. Department of Education, National Center for Education Statistics (2018). *Digest of Education Statistics, 2016* (NCES 2017-094),Table 330.10.

Take a moment and think about what this might mean. If they simply list the tuition, fees, room and board for all schools and take the arithmetic average (add them up and divide by total number of schools); then they don't take into account that only 2,000 students at Bowdoin College paid over $80,000 while over 30,000 students at U.T. Austin only paid $20,000. Just averaging $20,000 and $80,000 would be $50,000 for the two schools. But adding the total amount paid by all students and dividing by all students we get:

$$\frac{2,000 \times \$80,000 + 30,000 \times \$20,000}{2,000 + 30,000} = \frac{\$760,000,000}{32,000} = \$23,750$$

So the enrollment weighted average of $23,750 is heavily weighted toward the $20,000 by the 30,000 students paying this compared to the much smaller number (2,000) of students at Bowdoin College. The numbers of students, 30,000 and 2,000, are called *weights*.

As you read, jot down notes and questions. At the end of this section's Guided Worksheets there is a space for ***Reflection*** *and* ***Monitoring Your Understanding***, *where you can try and answer your questions after completing the guided activities and homework.*

Notes & Questions

Guided Practice Activity #1 – Weighted Grades

$$\textbf{Weighted Averages} = \frac{\text{weight}_1 \times \text{data}_1 + \text{weight}_2 \times \text{data}_2 + \cdots + \text{weight}_n \times \text{data}_n}{\text{sum of weights}}$$

1. You buy 8 children's tickets at $5.00 apiece and 2 adult tickets at $12.00 apiece.

    a. What is the average price per ticket?

    b. What are the *weights* in this weighted average problem?

2. Your grade consists of a test worth 80% and a paper worth 20%. You get a 70 on the test and a 100 on the paper.

    a. What grade do you get in the course?

    b. What are the *weights* in this problem?

3. Now assume there are two exams worth 40% each and a paper worth 20%. You get a 60 on the midterm and a 100 on the paper.

    a. What is your grade at this point in the course?

    b. What is the highest possible grade you can get in the class?

4. The following shows a gradebook for a class.

	A	B	C	D	E	F
1		HW Grades	Quiz Grades		Projects	
2		20%	20%		8% each	
3	1	7	7		93	
4	2	8	9			
5	3	6	7.5			
6	4	10	7.8			
7	5	9				
8	6					
9	7				Midterm	Final
10	8				10%	10%
11	9					
12	10					
13	11					
14	12				Grade:	
15						
16	Average	8	7.825			
17						

a. What formula is in cell **B16**?

b. What is this student's grade in the class right now?

c. What will her grade be if she gets a 100 on the next Project and a 95 on the Midterm?

Your Grade Point Average (GPA) is a statistic that most students are acutely aware of, but oddly cannot compute on their own. The GPA is a weighted average of your grades on a 4 point scale, weighted by the credits for each class. So an A in a 2 credit class contributes fewer quality points to your GPA than a B in a 4 credit class.

	A	B	C	D	E	F
1						
2						a.)
3		Classes	Credits	Grade	4 Point Scale	Quality Points
4		Calculus 1	4	A-	3.67	
5		Writing 1	4	C	2.00	
6		Theatre	3	B+	3.33	
7		Sociology	4	B	3.00	
8		Golf	2	D	1.00	
9	b.)	Total			Total	
10						
11			c.)	GPA:		
12						

1. The product of credits and grade on a 4 point scale is called quality points. Compute the quality points for an A in a 2 credit class and a B in a 3 credit class.

2. Compute the GPA for an A in a 2 credit class and a B in a 3 credit class.

3. Compute the GPA for the courses in the spreadsheet.

	A	B	C	D	E	F
1						
2						a.)
3		Classes	Credits	Grade	4 Point Scale	Quality Points
4		Calculus 1	4	A-	3.67	14.7
5		Writing 1	4	C	2.00	8.0
6		Theatre	3	B+	3.33	10.0
7		Sociology	4	B	3.00	12.0
8		Golf	2	D	1.00	2.0
9	b.)	Total	17		Total	46.7
10						
11			c.)	GPA:	2.75	
12						

Can you think of a new situation that would require the use of weighted averages besides GPA and grades?

Most students have never used weighted averages before. Do you think they are important enough to be taught? Why or why not?

Section 2.3 Proportionality

| **Objective 1** – Understand Proportionality |
| **Objective 2** – Compute the Constant of Proportionality |
| **Objective 3:** – Use Equations $y = k \cdot x$ |

| **Objective 1** – Understand Proportionality |

Key Terms

Proportional or in proportion	

Summary

In Chapter 1 we introduced the concept of **relationships** between inputs and outputs. In Chapter 2 we have introduced the basic idea of **comparing** the relative size of two quantities. Sometimes the two quantities we are comparing have a special relationship, doubling the size of one quantity necessitates a doubling of the size of other related quantity. The simplest example is the weight of water at 8 pounds per gallon. The weight will double if I double the volume (number of gallons): 2 gallons weigh 16 pounds and 3 gallons weigh 24 pounds. Contrast this with the number of gallons of water in your home and the number of soda cans. We can certainly compare the size of these quantities, and maybe get the ratio 2 gallons : 15 cans; but now doubling the number of gallons (buying two more gallons) does not mean we also have double the number of cans of soda, the quantities are not linked in this special way of doubling.

Quantities that are linked in this special way of doubling one means the other also must double are said to be proportional.

*As you read, jot down notes and questions. At the end of this section's Guided Worksheets there is a space for **Reflection** and **Monitoring Your Understanding**, where you can try and answer your questions after completing the guided activities and homework.*

Notes & Questions

Guided Practice Activity #1 – Proportionality

- Determine which of the following quantities are proportional.

The first one has been done for you:

1. Property tax : Appraised value of your house

   - YES proportional, doubling value will double the property tax.

2. population of a region : area of the region

3. volume of gold : weight of gold

4. population of NY City : population of NY state

> **Intuitive Definition:** Two quantities are **directly proportional** if when you double one quantity …

5. number of babies born in a year : number of women aged 12-45.

6. the total cost of buying tickets : number of tickets

7. height of a digital photo : width of a digital photo

8. the number of accidental deaths in 2005 : the number of motor vehicle deaths

9. cups of sugar : cups of flour (in a pancake recipe)

10. your IQ : your GPA

11. distance to work : time it takes to get to work

12. distance to a celestial object : speed at which it is receding from us

13. euros : dollars (at an exchange booth)

14. sea level change : amount of melted ice from glaciers and icebergs

15. miles you drive : gallons of gas used

Key Terms

Constant of proportionality	

Summary

In objective 1 of this section we identified a special relationship that can occur when comparing two quantities. In essence, proportionality implies the "ratio is constant". The weight of water is proportional to its volume:

- 8 pounds : 1 gallon
- 16 pounds : 2 gallons
- 24 pounds : 3 gallons

All of these ratios have the same 8 : 1 comparison and are thus "constant". We typically represent ratios vertically like fractions and now get the idea of equivalent fractions:

$$\frac{8}{1} = \frac{16}{2} = \frac{24}{3} = 8$$

The ratios "disappear" into the constant…

Ratios have order but there is usually nothing special about the order chosen.

- 1 gallon : 8 pounds
- 2 gallons : 16 pounds
- 3 gallons : 24 pounds

This order leads to these equivalent fractions (reciprocals of the above fractions) and a different constant:

$$\frac{1}{8} = \frac{2}{16} = \frac{3}{24} = 0.125$$

Thus proportionality implies a constant ratio, but leads to two possible constants that are reciprocals.

As you read, jot down notes and questions. At the end of this section's Guided Worksheets there is a space for **Reflection** *and* **Monitoring Your Understanding***, where you can try and answer your questions after completing the guided activities and homework.*

Notes & Questions

Guided Practice Activity #1 – Constants of Proportionality

- Determine which of the following quantities are proportional.
- If they are proportional, determine the constant of proportionality.
- Specify the units of the constant of proportionality, and identify the name of the constant if you know it.

The first one has been done for you:

1. Property tax ($2,000) : Appraised value of your house ($180,000).

   - YES proportional
   - $$k = \frac{\text{Tax \$}}{\text{House value \$}} = \frac{\$2,000}{\$180,000} = 0.111$$
   - These units cancel, so this constant of proportionality is unit-less. It is called the tax rate or the mill rate, and is often reported as tax per $1,000 of value: $11.11 in this case.

2. population of a region (4.5 million) : area of the region (77,000 square miles)

3. volume of water (5 gallons): weight of water (40 pounds)

4. population of NY City (8.6 million) : population of NY state (19.85 million)

> **Intuitive Definition:** Two quantities are **directly proportional** if when you double one quantity …
>
>

5. number of babies born in a year (2.4 million) : number of women aged 12-45 (110 million)

6. the total cost of buying tickets ($24): number of tickets (6)

7. height of a digital photo (3 inches) : width of a digital photo (5 inches)

8. the number of accidental deaths in 2005 (120,000): the number of motor vehicle deaths (43,510)

9. cups of sugar (1): cups of flour (3) (in a pancake recipe)

10. your IQ (115): your GPA (3.4)

11. distance to work (12 miles): time it takes to get to work (1/2 hour)

12. distance to a celestial object (5 megaparsecs): speed at which it is receding from us (2500 km/s)

13. euros (80): dollars (100) (at an exchange booth)

14. sea level change (150 mm): amount of melted ice from glaciers and icebergs (1 gazillion gallons)

15. miles you drive (120): gallons of gas used (4)

Summary

Given two proportional quantities we can represent their relationship with an equation. The volume of water is proportional to its weight with the constant ratio, 1 gallon : 8 pounds. Any volume (gal), v, will have an associated weight (lbs), w, and the ratio of the two quantities will be 1 : 8. This leads to setting up a proportion with **both variables**:

$$\frac{v}{w} = \frac{1}{8}$$

Cross multiplying gives an **equation**:

$$8 \cdot v = 1 \cdot w$$

We can solve this equation for either variable, giving two equivalent equations of the form, $y = k \cdot x$:

$$w = \frac{8}{1} \cdot v \qquad v = \frac{1}{8} \cdot w$$

Reducing the fraction to a decimal gives us two different looking equations, the reciprocality of the constants is hidden:

$$w = 8 \cdot v \qquad v = 0.125 \cdot w$$

CAUTION! The ratio 1 gal : 8 lbs involves units (gal and lbs), while the equation involves variables w = number of pounds and v = number of gallons. In the ratio, the 8 is in front of the unit for weight (lbs), but in the equation the 8 switches sides and is front of the variable for volume.

Ratio		Equation
1 gal : 8 lbs	versus	$w = 8 \cdot v$
Units (gal and lbs)		Variables (w and v)
Arithmetic		Algebra!

As you read, jot down notes and questions. At the end of this section's Guided Worksheets there is a space for **Reflection** *and* **Monitoring Your Understanding***, where you can try and answer your questions after completing the guided activities and homework.*

Notes & Questions

Guided Practice Activity #1 – Constants of Proportionality

Given the following proportional quantities, determine the two equations, $y = k \cdot x$, representing the relationship. The first one has been done for you:

1. Property tax ($2,000) : Appraised value of your house ($180,000)

$$\frac{T}{V} = \frac{\$20,000}{\$180,000} = \frac{1}{9}$$
$$9 \cdot T = 1 \cdot V$$
$$V = 9 \cdot T \qquad T = \frac{1}{9} \cdot V$$

2. euros (113) : British pounds (100)

3. the total cost of buying tickets ($24): number of tickets (6)

4. height of a digital photo (3 inches) : width of a digital photo (5 inches)

Excel Preview – 2.1 Currency

The following screenshot gives the ratio of three currencies, EUR : USD : YEN.

◢	A	B	C	D	E	F	G	H
1		Ratio of Currencies						
2		Euros	Dollars	Yen				
3		€ 3.00	$ 4.00	¥ 500.00				
4								
5	Input			Input			Input	
6	Dollars:	$ 200.00		Yen:	¥4,000.00		Euros:	€ 60.00
7		a.)			a.)			a.)
8	Outputs			Outputs			Outputs	
9	Euros:			Euros:			Dollars:	
10	Yen:			Dollars:			Yen:	
11								

1. Convert $200 into euros and yen.

2. Find equations for euros as a function of dollars and yen as a function of dollars.

3. What formulas are entered in cells B9 and B10?

Excel Preview – 2.2 Eddie B

A clothing retailer gives you 10 points for every $1 spent, and you can trade in 2,550 points for a $15 gift certificate.

1. How much do you have to spend to get one gift certificate?

2. What is the value of the gift certificate?

3. Why are the two quantities, what you spend and the value of the associated gift certificates, proportional? (Assume fractional gift certificates are possible.)

4. Determine the equation, $y = k \cdot x$, representing the relationship between these two quantities with what you spend (*s*) as the input, and value of gift certificates (*v*) as output.

5. In the spreadsheet fractional gift certificates are not possible. You are instructed to use the **INT** function (which rounds any decimal down) to determine the whole number of gift certificates in cell **F17**. Determine the formula you enter in cell **F17**.

	A	B	C	D	E	F
1						
2		Eddie Bauer Shopping Spree			a.)	
3		Items	Quantity	Cost	Total Cost	Points
4		1 Polo Shirt	3	$ 50.00		
5		2 Dress Shirt	4	$ 75.00		
6		3 Khakis	2	$ 59.99		
7		4 Merrell Shoes	1	$ 135.00		
8		5 Snowline Jacket	1	$ 219.00		
9		6 Belt	1	$ 19.99		
10		7 Cloud Layer Fleece	2	$ 75.00		
11		8 Sweater	3	$ 80.00		
12		9 Full Zip Hoodie	1	$ 70.00		
13		10 Wrinkle Free Oxford	5	$ 55.00		
14						
15		Totals:	b.)			
16						
17			c.)	Gift Certificates:		
18				Value of Gift Certificates:		
19						

Section 2.3 Reflections
Monitor Your Understanding

What is one important tip you learned for determining the equation if two quantities are proportional?

Section 2.4 Financial Literacy

Objective 1 – Understand PE Ratios and Stock Quotes
Objective 2 – Understand Inflation and the CPI
Objective 3 – Compute Money Ratios for Retirement

Objective 1 – Understand PE Ratios and Stock Quotes

Key Terms

PE ratio	
Stock	
Dividends	
Share	
Price per share	
Retirement plan	
Quarter	
Earnings	
Profit	
Earnings per share	
Stock quote	

Summary

A share of stock in a company refers to a part ownership in the company. You can thus become part owner of any publicly traded company by buying shares using a brokerage account like TD Ameritrade or e-trade. On October 22, 2018 you could have bought one share of Apple

Incorporated for $220.65 at the close of the trading day (4:00 PM EST), or one share of Amazon for $1,789.30. Why the big difference in price?

There are several reasons:
1. Every company has a different number of shares for purchase: 487.7 million shares of Amazon and 4.8 billion shares of Apple. So fewer shares can contribute to higher price.
2. Every company makes a different amount of profit each year. This is reported as earnings per share: $11.53 earnings per share (EPS) for Apple and $11.01 earnings per share for Amazon.
3. Finally, every company has different projections for growth. A more robust growth outlook will lead to a higher share price now.

Companies exist to make money, (i.e. profit). This may sound strange to you. Apple makes phones and computers, but if they are not profitable they will go out of business, so profit is the number one priority. Thus Apple's profit of $11.53 per share makes its shares more valuable than Amazon's shares at $11.01 EPS. Yet Amazon's stock price is much higher! This discrepancy is captured in the PE ratio, which is the ratio of share price to EPS.

Price to Earnings (PE) Ratios

Apple	$\dfrac{\$220.65}{\$11.53} = 19.14$
Amazon	$\dfrac{\$1,789.30}{\$11.01} = 162.52$

People are currently paying 162.5 times what Amazon earns per share versus only 19.1 times for Apple! In this sense, the share price of Amazon is much more expensive. Obviously people think Amazon has huge growth potential to pay this much its shares.

As you read, jot down notes and questions. At the end of this section's Guided Worksheets there is a space for **Reflection** *and* **Monitoring Your Understanding**, *where you can try and answer your questions after completing the guided activities and homework.*

Notes & Questions

Guided Practice Activity #1 – PE ratios

The Price to Earnings, or PE ratio, is one of the most common statistics used to assess the "value" of a share of stock. Stocks will be fully discussed in Chapter 9, but for now we want to introduce the basic idea of the PE ratio.

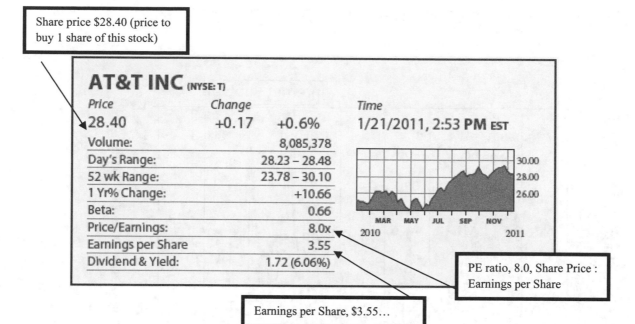

Share price $28.40 (price to buy 1 share of this stock)

PE ratio, 8.0, Share Price : Earnings per Share

Earnings per Share, $3.55…

1. Compute the ratio of the price per share to the earnings per share and scale the second quantity to 1.

2. The earnings per share (EPS) statistic is the ratio of the profits the company made for the previous 12 months to the number of the shares available. If there were 5.9 billion shares available, use the EPS statistic to compute the profits that AT&T made.

3. The share price changes constantly as people trade shares with each other. If there are more buyers than sellers, the price goes up. If there are more sellers than buyers the share price goes down. What are two different ways the PE ratio can increase?

4. Compute the ratio of the dividend to the price per share, and scale the second quantity to 100.

5. The following graphic found in Peter Schiller's book, Irrational Exuberance, displays both PE ratios for the entire stock market and long term interest rates. What would you estimate is the average historical PE ratio?

6. If PE ratios rise, what seems to happen to long term interest rates?

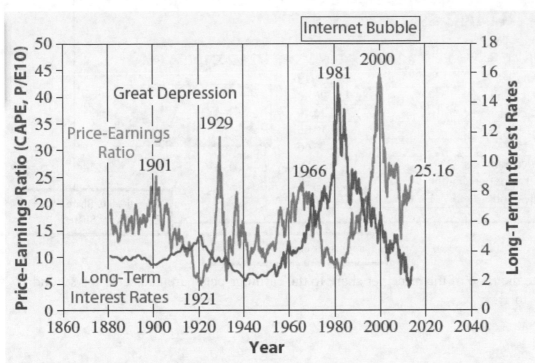

FIGURE 2.8 Schiller's Graphic, Showing Historical PE Ratios Against Interest Rates
Data from: *Irrational Exuberance* by Robert J. Shiller

Objective 2 – Understand Inflation and the CPI

Key Terms

Consumer price index	
Inflation	

Summary

Inflation is an odd thing. Why do prices keep going up? As populations grow we would expect there to be less money per person floating around ("in circulation"), so obviously the government must print more money. If they double the amount (value) of money in circulation, then every dollar will lose half its value, or everything will cost twice as much. The Federal Reserve website[6] tells us: "There was approximately $1.69 trillion in circulation as of September 26, 2018, of which $1.64 trillion was in Federal Reserve notes." The Treasury Department website[7] states: "During Fiscal Year 2014, the Bureau of Engraving and Printing delivered approximately 6.6 billion notes to the Federal Reserve, producing approximately 24.8 million notes a day with a face value of approximately $560 million." So $560 million a day is printed, most of which (~95%) goes to replacing worn out bills, the other 5%, $28 million per day, is new money put into circulation! Prices then rise naturally (inflation) in response to the extra cash available.

The Consumer Price Index tracks the rise in prices by setting the value of a basket of goods (loaf of bread, head of lettuce, etc.) to 100 in 1983-84. The CPI then measures the price of this basket of goods for every subsequent year; in 2013 these same goods would have cost $233, meaning prices had more than doubled since 1983.

*As you read, jot down notes and questions. At the end of this section's Guided Worksheets there is a space for **Reflection** and **Monitoring Your Understanding**, where you can try and answer your questions after completing the guided activities and homework.*

Notes & Questions

---

[6] https://www.federalreserve.gov/faqs/currency_12773.htm
[7] https://www.treasury.gov/resource-center/faqs/Currency/Pages/edu_faq_currency_production.aspx

Guided Practice Activity #1 – PE ratios

The *Consumer Price Index* (**CPI**) is a financial ratio that has to do with inflation. The costs of goods and services typically rise each year. For example, a movie ticket in 1990 that cost $6.50 will cost more in 2005. Thus a dollar in 1990 (1990$) was worth **more** than a dollar in 2005 (2005$).

It is helpful to think of 1990$ and 2005$ as different currency in much the same way that dollars and euros are different currency. We need to convert between these currencies and the CPI allows us to do that.

The CPI is a measure of the cost of a fixed basket of goods and services in a given year. This is called an **index** because it has arbitrarily been set to 100 for the cost of this basket in 1983$. These same goods and services would have then cost 130.7 in 1990$ and 195.3 in 2005$.

To convert the cost of a $6.50 movie ticket in 1990 to 2005$ we can set up a proportion using the CPI values:

	CPI	Cost of an Item
1990$	130.7	6.50
2005$	195.3	

1. Set up a proportion and compute the cost of the movie ticket in 2005$ which originally cost $6.50 in 1990$.

2. Assume the ratio of euros to dollars is 4€ : 5$. What is more expensive, a mocha latte that costs 2€ or one that costs $3.50?

		Cost of an Item	
	Conversion	2€ latte	$3.50 latte
EUR	4	2	
USD	5	x	3.5

3. Which was more expensive: a $2 Starbucks mocha latte in 1990 or a $3.50 Starbucks mocha latte in 2005?

Cost of an Item

	CPI	1990	2005
1990$	130.7	2	
2005$	195.3	x	3.5

The CPI allows us to compare datasets that involve monetary values over time. The amount of money the Federal Government collected (i.e. receipts (taxes)) looks to have been increasing steadily in the table below. But in order to compare these values we need to convert all three into **one currency**.

		1990	2000	2005
	CPI	US Receipts (billions)		
1990$	130.7	**1,032**		
2000$	172.2		**2,025**	
2005$	195.3			**2,154**

4. Convert the 3 receipt values in the table to **one currency** using the CPI values (you can choose which currency to use), and determine the year when the receipts are greatest when adjusted for inflation.

		1990	2000	2005
	CPI	US Receipts (billions)		
1990$	130.7	**1,032**		
2000$	172.2		**2,025**	
2005$	195.3	a	b	**2,154**

Summary

Financial planning involves saving money for retirement and paying off loans for education and home ownership (mortgages). In this table you are given money ratios from the book, *Your Money Ratios*, comparing amounts related to your savings and loans. Income and earnings refer to your annual salary. These are just suggested guidelines meant to get you thinking about these topics, the stock : bond suggested allocation in particular varies widely depending on who you talk to.

Your Age	Capital : Income	Savings : Income	Mortgage : Income	Education: Earnings	Stock : Bond
25	0.1	12%	2.0	0.75	50% : 50%
30	0.6	12%	2.0	0.45	50% : 50%
35	1.4	12%	1.9	0.00	50% : 50%
40	2.4	12%	1.8	X	50% : 50%
45	3.7	15%	1.7	X	50% : 50%
50	5.2	15%	1.5	X	50% : 50%
55	7.1	15%	1.2	X	50% : 50%
60	9.4	15%	0.7	X	40% : 60%
65	12.0	15%	0.0	X	40% : 60%

As you read, jot down notes and questions. At the end of this section's Guided Worksheets there is a space for **Reflection** *and* **Monitoring Your Understanding***, where you can try and answer your questions after completing the guided activities and homework.*

Notes & Questions

Guided Practice Activity #1 – Money Ratios

1. You are 35 years old and make $60,000. How much *Capital* should you have?

2. How much should you be saving this year?

3. How much should the balance on your mortgage be?

4. What should you owe on student loans?

5. How much of your capital should be invested in bonds?

6. By the time you are 65 what salary do you anticipate making? How much capital should you have at that point for retirement?

Excel Preview – 2.4 CPI

This spreadsheet is giving CPI values (annual and monthly).

	I	J	K	L	M	N	O	P	Q
1		This problem uses Consumer Price Index (CPI) data from the U.S. Bureau of Labor Statistics							
2		(BLS) from 1913 to 2017. Note the years may be hidden, scroll to go all the way back to							
3		1913.							
4									
5									
6									
7									
8									
9							a.)	b.)	c.)
10								Ratio	
11	Aug	Sep	Oct	Nov	Dec	Annual	Cost of iPhone 8	This Year CPI : Previous Year CPI	Annual Inflation Rate
112	233.9	234.1	233.5	233.1	233.0	233.0			
113	237.9	238.0	237.4	236.2	234.8	236.7			
114	238.3	237.9	237.8	237.3	236.5	237.0			
115	240.8	241.4	241.7	241.4	241.4	240.0			
116	245.5	246.8	246.7	246.7	246.5	245.1	$ 699.00		
117									

1. Assuming an iphone 8 cost $699 in 2017 (row 116), use the CPI values to compute how much it would have cost in 2016$.

2. Next compute how much it would have cost in 2015$. There are two ways to do this make sure to identify both ways.

3. What formula could you type in cell O115 that you can fill up?

4. To compute the inflation rate we take the ratio of a given year's CPI to the previous year, subtract 1 , and interpret the result as a percentage (this is the percent change discussed in Chapter 4). Compute the inflation rate for 2017.

Excel Preview – 2.8 Receipts

	A	B	C	D	E	F	G	H	I	J	K	L
1	Table 1.3 - SUMMARY OF RECEIPTS, OUTLAYS, AND SURPLUSES OR DEFICITS (-) IN CURRENT DOLLARS, CONSTANT (FY 2009) DOLLARS, AND AS PERCENTAGES OF GDP: 1940 - 2023											
2						(dollar amounts in billions)						
3		In Current Dollars			In Constant (FY 2009) Dollars				As Percentages of GDP			
4	Fiscal Year	Receipts	Outlays	Surplus or Deficit (-)	Receipts	Outlays	Surplus or Deficit (-)	Addendum: Composite Deflator	Receipts	Outlays	Surplus or Deficit (-)	
72	2006	2,406.9	2,655.0	-248.2	2,561.9	2,826.0	-264.2	0.9395	17.6	19.4	-1.8	
73	2007	2,568.0	2,728.7	-160.7	2,663.1	2,829.7	-166.7	0.9643	17.9	19.1	-1.1	
74	2008	2,524.0	2,982.5	-458.6	2,529.0	2,988.5	-459.5	0.9980	17.1	20.2	-3.1	
75	2009	2,105.0	3,517.7	-1,412.7	2,105.0	3,517.7	-1,412.7	1.0000	14.6	24.4	-9.8	
76	2010	2,162.7	3,457.1	-1,294.4	2,129.3	3,403.6	-1,274.4	1.0157	14.6	23.4	-8.7	
77	2011	2,303.5	3,603.1	-1,299.6	2,215.9	3,466.2	-1,250.2	1.0395	15.0	23.4	-8.5	
78	2012	2,450.0	3,536.9	-1,087.0	2,310.7	3,335.8	-1,025.1	1.0603	15.3	22.1	-6.8	
79	2013	2,775.1	3,454.6	-679.5	2,583.9	3,216.6	-632.7	1.0740	16.8	20.9	-4.1	
80	2014	3,021.5	3,506.1	-484.6	2,770.7	3,215.1	-444.4	1.0905	17.5	20.3	-2.8	
81	2015	3,249.9	3,688.4	-438.5	2,962.0	3,361.6	-399.7	1.0972	18.1	20.5	-2.4	
82	2016	3,268.0	3,852.6	-584.7	2,953.4	3,481.8	-528.4	1.1065	17.7	20.9	-3.2	
83	2017	3,316.2	3,981.6	-665.4	2,940.7	3,530.7	-590.0	1.1277	17.3	20.8	-3.5	
84	2018 estimate	3,340.4	4,173.0	-832.6	2,900.6	3,623.6	-723.0	1.1516	16.7	20.8	-4.2	
85	2019 estimate	3,422.3	4,406.7	-984.4	2,917.8	3,757.1	-839.3	1.1729	16.3	21.0	-4.7	
86	2020 estimate	3,608.9	4,595.9	-986.9	3,015.7	3,840.5	-824.7	1.1967	16.4	20.8	-4.5	

1. What are receipts and outlays?

2. Which column of receipts have been adjusted for inflation so we can accurately compare them: Current Dollars or Constant (Fiscal Year 2009) Dollars?

3. How much money did the government collect in 2017?

4. Why is the surplus/deficit column always negative?

5. The Receipt value of $3,316.2 in 2017 was adjusted to 2,940.7 in 2009$. The CPI in 2009 was 214.5, compute the CPI in 2017 using these numbers.

| Objective 1 – Compute z-scores and Standardized Scores |
| Objective 2 – Explore the Normal Distribution |

| Objective 1 – Compute z-scores and Standardized Scores |

Key Terms

z-scores	
Chebychev's Theorem:	

Summary

In the last chapter's *Spotlight on Statistics* we introduced the basic measures of central tendency and spread. The mean is the arithmetical average of the data set and the standard deviation is the "average distance from the mean." Given these two descriptive statistics we can put into perspective the location of individual data points within the overall distribution of values. As explained in the beginning of this chapter, absolute numbers are virtually meaningless without comparison.

The age distribution of the U.S. shows a mean age 38.5 in 2016 with a standard deviation of 23.1 years. Thus someone who is 15 years old will be about one standard deviation below the mean and someone who is 62 is one standard deviation above the mean. These ages represent the average distance away from the mean. One of the fundamental statistical questions we will ask is whether or not a data value is **unusual**. A z-score will tell how many standard deviations away from the mean a given data value is.

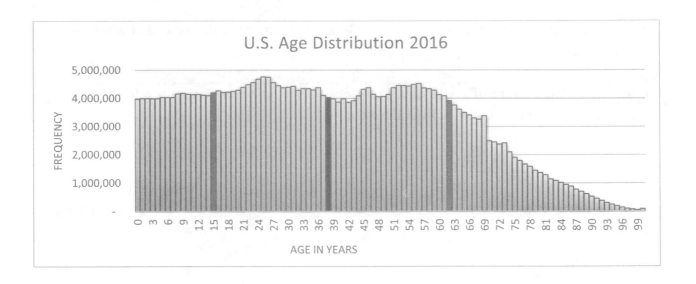

As you read, jot down notes and questions. At the end of this section's Guided Worksheets there is a space for **Reflection** and **Monitoring Your Understanding**, where you can try and answer your questions after completing the guided activities and homework.

Notes & Questions

Guided Practice Activity #1 – *z*-scores

You get a 63 on a test. **Is this bad?**

To answer this question we need to know how this score compares to the rest of the class. Assuming the mean was 51 we now know you scored above average. **But how far above average?**

To answer this question we need to know how spread out the scores are. If all the other scores are between 49 and 53 then you did extremely well (A+), but if a bunch of people got in the 90's and a bunch got in the 20's then you just did OK (B-).

The **standard deviation** (average distance from the mean) gives us a measure of this spread. Assuming it was 5 points, your 63 is more than twice that distance above the mean.

$$\text{Mean } \overline{X} = 51 \qquad \text{StDev } s = 5$$

$$z = \frac{63 - 51}{5} = \frac{12}{5} = 2.4$$

In fact your 63 is 12 points above the mean and the distance of 12 is 2.4 times the average distance of 5 points. The number +2.4 is your **z-score**: it tells me you are 2.4 times the standard deviation from the mean. Someone who got a 39 is 12 points below the mean and would have a z-score of -2.4.

Definition: The **z-score** of a data value is the ratio of the data value's distance from the mean to the average distance from the mean:

$$z = \frac{x - \overline{X}}{s}$$

---

1. If you get a 63 and the mean is $\overline{X} = 51$ and the standard deviation is $SD = 2$, what is your z-score?

2. Z-scores allow you to compare different scales. For example, the admissions office is comparing two different applicants, one has a math SAT score of 640 and the other has a math ACT score of 28. Which student has a better score? You need to know that for the SAT: $\overline{X} = 500$ and $s = 100$, and for the ACT: $\overline{X} = 22$ and $s = 5$.

3. The following table shows a dataset {69, 72, 75, 78, 81} with a mean of 75 and standard deviation of 4.74. Fill in the rest of the table using the formulas in the first column (some values have been computed for you):

	Data Values					Mean	StDev
Raw Scores	69	72	75	78	81	75	4.74
Z-scores		-0.63			1.27	0	1
T-scores = 50+10*z					62.7	50	10
IQ = 100 + 15*z	80.95						
SAT = 500 + 100*z				563			

	Data Values					Mean	StDev
Raw Scores	69	72	75	78	81	75	4.74
Z-scores	☐	-0.63	☐	☐	1.27	0	1
T-scores = 50+10*z	☐	☐	☐	☐	62.7	50	10
IQ = 100 + 15*z	80.95	☐	☐	☐	☐	☐	☐
SAT = 500 + 100*z	☐	☐	☐	563	☐	☐	☐

4. In the example above with a mean of 75 and standard deviation of 4.74, what raw score would have a z-score of $+1$? -1? $+2$? -2?

5. A graduate school grades with the following system based on z-scores. If a class of 10 students takes a quiz and half get a 90 and half get a 92, the mean is 91 and the standard deviation is 1.05. What are their letter grades using this system?

A	$z > 2$
B	$1 \le z < 2$
C	$z < 1$

6. Do you like or dislike this system? Explain!

Key Terms

Bell shaped curve	
Normal Distribution	
95% Rule	
Fat tail	

Summary

The bell shaped symmetric curve is an approximation for many real world data sets (heights of people, housing prices, stock market price changes,…). Given the variety of possible units we replace our **data** values with **z-scores**, and talk about how many **standard deviations** away from the **mean** a value is.

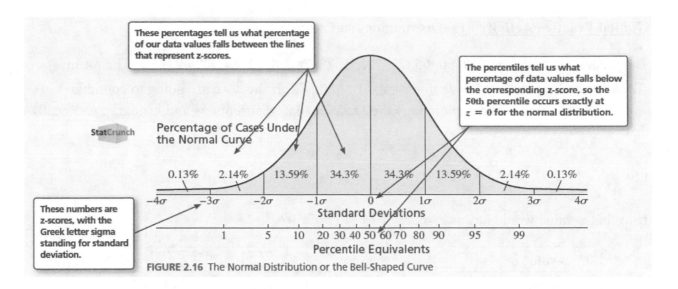

These percentages tell us what percentage of our data values falls between the lines that represent z-scores.

The percentiles tell us what percentage of data values falls below the corresponding z-score, so the 50th percentile occurs exactly at $z = 0$ for the normal distribution.

Percentage of Cases Under the Normal Curve

These numbers are z-scores, with the Greek letter sigma standing for standard deviation.

| 0.13% | 2.14% | 13.59% | 34.3% | 34.3% | 13.59% | 2.14% | 0.13% |

Standard Deviations

Percentile Equivalents

FIGURE 2.16 The Normal Distribution or the Bell-Shaped Curve

Chebychev's Theorem told us that in any distribution 75% of the data would be within two standard deviations of the mean. We can see that for a normal distribution we get 95% of our data within two standard deviations of the mean. Having a z-score greater than 2 or less than -2 is considered **unusual**.

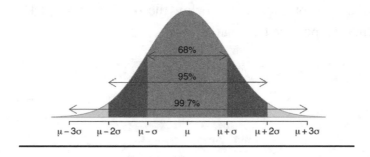

*As you read, jot down notes and questions. At the end of this section's Guided Worksheets there is a space for **Reflection** and **Monitoring Your Understanding**, where you can try and answer your questions after completing the guided activities and homework.*

Notes & Questions

<u>**Guided Practice Activity #1**</u> – Distributions and z-scores

The grades in a class are {40, 60, 65, 72, 72, 75, 78, 78, 78, 83, 84, 87, 92, 95}. The mean $\overline{X} = 75.6$ and the standard deviation $s = 14.1$. In this worksheet we are going to continue working with z-scores and explore how they relate to data distributions and histograms. Recall that:

$$z = \frac{x - \overline{X}}{s}$$

1. What x-value would have a z-score of +1? –1? –2? 0?

x-value				
z-score	-2	-1	0	+1

2. Using the dot plot below, put tick marks below the x-axis at $z = 0, \pm 1, -2$ and write these numbers below their respective tick marks.

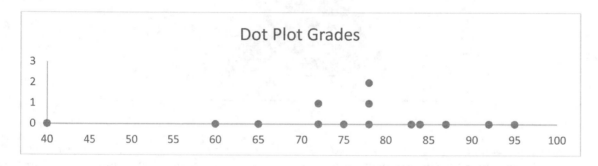

Now we are going to draw a **Histogram** of the data. First you must choose intervals or "bins" into which the data values will fall. With grades it is typical for the bins to be intervals of length 10.

3. Count the number of data values in each bin (these are called the frequencies) and fill in the table.

Frequency						
Bins	40-49	50-59	60-69	70-79	80-89	90-99

4. Now draw a histogram (column chart) next to the table with bins on the x-axis and the frequency counts as heights of the columns.

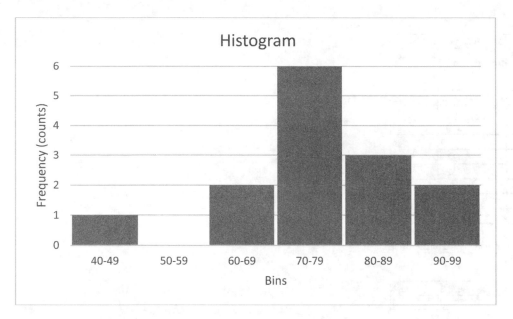

5. Why is the histogram referred to as a "distribution" of the data values?

Consider the normal distribution shown below with the classic bell shaped curve. It is very important you understand that this is a histogram like the one you just drew. The bins are very small, meaning the columns are very narrow, giving the impression of a smooth curve. Many datasets, such as IQ-scores and SAT scores, will have this distribution or shape.

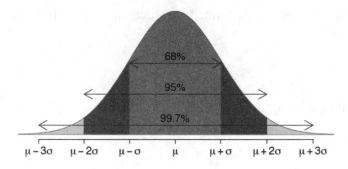

6. If IQ scores have a mean of 100 with a standard deviation of 15, 95% of people will have IQ scores between what two values?

◢	A	B	C	D	E
1				a.)	
2		Planet	Orbital Radius	Ratio	
3			(millions km)	Planet to Earth	
4		Mercury	57.9	0.387	
5		Venus	108.2	0.723	
6		Earth	149.6	1.000	
7		Mars	227.9	1.523	
8		Asteroids	686.1	4.586	
9		Jupiter	778.3	5.203	
10		Saturn	1429.4	9.555	
11		Uranus	2875	19.218	
12		Neptune	4504.5	30.110	
13		Pluto	5915.8	39.544	
14					
15		Pluto is how many times farther than Earth			
16	b.)	39.5			
17					
18	c.)	Mean:	1673.3	11.2	
19	c.)	StDev:	2075.3	13.9	
20					

Notice that the distance of each planet from the sun has been scaled with earth's distance set to 1 (this distance is called an astronomical unit (AU)). The mean and standard deviations for the orbital radii in both millions of kilometers and AU's are given.

1. How many AU's is Venus from the sun?

2. Compute the z-score for Earth's distance from the sun in both millions of kilometers and AU's.

3. Why doesn't it matter what units you use?

4. Compute the T-score for Earth.

5. How does a T-score indicate below the mean?

Monitor Your Understanding

1. Was it difficult for you to understand standardized scores? Why or why not?

2. What is meant by the "shape" of your data?

Chapter 3 – Units, Conversions, Scales, and Rates

Section 3.1 Units and Conversions

Objective 1 – Understand and Convert Units
Objective 2 – Use Dimensional Analysis
Objective 3 – Compare Ratio and Interval Units

Objective 1 – Understand and Convert Units

Key Terms

Units	
Units of observation	
Units of measurement	
Conversion	
Conversion factor	

Summary

The act of measuring a quantity requires us to decide what units or system of measurement will be used. When we say a quantity is 4, the units answer the question: "Four what?" The key distinction between units of observation (nationality, gender, etc.) and units of measurement (height, volume, etc.) is that we can always convert between different choices of units of measurement. Given any unit of length we can convert this to any other unit of length, whereas we cannot convert a category of people into another category. Different choices of units of measurement (cm, ft, etc.) are typically proportional to each other. Conversions therefore are perfect applications of our ratio and proportion material from chapter 2.

Notes & Questions

Guided Practice Activity #1 – Length Conversions

Conversions are ratios between units of measurement. The following units of length are all proportional.

1. For each of the following conversions, write down the equation representing the proportional relationship. Be sure to use a variable to represent the unit, like F for feet, don't use the units themselves (like ft). The first one has been done for you.

Length Conversion Table			$y = k \cdot x$
1 centimeter (cm)	=	10 millimeters (mm)	$m = 10 \cdot C$
1 inch	=	2.54 centimeters (cm)	
1 foot	=	0.3048 meters (m)	
1 foot	=	12 inches	
1 yard	=	3 feet	
1 meter (m)	=	100 centimeters (cm)	
1 meter (m)	≈	3.281 feet	
1 furlong	=	660 feet	
1 kilometer (km)	=	1000 meters (m)	
1 kilometer (km)	≈	0.6214 miles	
1 mile	=	5280 ft	
1 mile	=	1.609 kilometers (km)	
1 nautical mile	=	1.852 kilometers (km)	

2. Convert the following (2 decimal place accuracy unless exact answer):

   a. 6 ft to m

   b. 64 in. to ft

c. 350 cm to ft

d. 149.6 km to mi

e. 1000 mm to m

f. 3 mi to m

g. 149.6 million km to million mi

h. 480 cm to mm

Key Terms

Dimensional Analysis	

Summary

Converting between units can be done using our tried and true technique of setting up a proportion and cross multiplying or by a new technique of canceling units called dimensional analysis. To convert 10 inches to centimeters using the conversion 1 in. : 2.54 cm we have two options.

$$\frac{10 \text{ in.}}{x \text{ cm}} = \frac{1 \text{ in.}}{2.54 \text{ cm}}$$

$$10 \times 2.54 = x \times 1$$

$$x = 10 \times 2.54 = 25.4 \text{ cm.}$$

$$10 \; \cancel{\text{in.}} \times \frac{2.54 \text{ cm}}{1 \; \cancel{\text{in.}}} = 25.4 \text{ cm}$$

Dimensional analysis is "simpler" in that it is one step, but conceptually it can be more abstract than the intuitive 10 inches is to x cm as 1 inch is to 2.54 cm.

As you read, jot down notes and questions. At the end of this section's Guided Worksheets there is a space for **Reflection** *and* **Monitoring Your Understanding**, *where you can try and answer your questions after completing the guided activities and homework.*

Notes & Questions

Guided Practice Activity #1 – Proportions vs. Dimensional Analysis

Convert the following using two techniques, setting up a proportion and dimensional analysis (1 decimal place accuracy unless exact answer).

1. Convert 82 feet to meters using 3.281 feet : 1 meter.

2. Convert 3 days to minutes using 24 hours : 1 day and 60 minutes : 1 hour.

3. Convert $250 to Russian rubles using 20 USD : 1340 RUB.

4. Convert 120 pounds to kilograms using 2.2 lbs : 1 kg.

5. Convert 40 liters to fluid ounces using 1000 ml : 1 liter and 29.57 ml : 1 fl. oz.

Objective 3 – Compare Ratio and Interval Units

Key Terms

Ratio units	
Interval units	

Summary

There do exist units for the same quantity which are not proportional, which may seem odd. The best example is units for temperature. Doubling the number of degrees Fahrenheit does not correspond to a doubling of degrees Celsius. Doubling does not work for these units because they do not have a well-defined zero, which represents the absence of something. Zero degrees Fahrenheit was chosen to represent the freezing point of a brine solution, a totally arbitrary choice of zero; while 0°C corresponds to the freezing point of water. Units like feet and dollars have a well-defined zero that represents the absence of the underlying quantity (no length or money respectively).

As you read, jot down notes and questions. At the end of this section's Guided Worksheets there is a space for **Reflection** *and* **Monitoring Your Understanding***, where you can try and answer your questions after completing the guided activities and homework.*

Notes & Questions

Guided Practice Activity #1 – Temperatures

1. Using the equation convert the following temperatures (2 decimal place accuracy unless exact answer):

$$C = \frac{5}{9} \cdot F - \frac{160}{9}$$

a. 40°F to Celsius

b. 80°F to Celsius

c. Did doubling the degrees Fahrenheit result in doubling the degrees Celsius?

d. 0°F to Celsius

e. Does 0°F correspond to to 0°C?

f. 50°C to Fahrenheit

g. 25°C to Fahrenheit

h. Solve the equation for F as a function of C.

Excel Preview – 3.1 Conversions

	A	B	C	D	E	F	G
1	Convert units for length, mass, volume, and speed using the given ratio						
2	tables. In particular, use mixed and absolute cell references so that your						
3	formulas are able to be filled across each row. Format all answers to 2						
4	decimal places.						
5							
6							
7			Ratio Tables				
8	in	ft	yd	cm	m		
9	36	3	1	91.44	0.9144		
10							

OUTPUTS

INPUTS		inches	feet	yards	centimeters	meters
inches	12	☐	☐	☐	☐	☐
feet	3	☐	☐	☐	☐	☐
yards	1	☐	☐	☐	☐	☐
centimeters	100	☐	☐	☐	☐	☐
meters	1	☐	☐	☐	☐	☐

1. Compute the conversions for the table using the inputs given and the ratio table with the conversions above.

2. The first output cell for inches is in cell **K3**. What formula can you enter in this cell that can be **filled across** using the ratio table above in row 9 and the input cell **J3**? Hint: It might be easier to determine the formula in cell **L3** for feet first.

Section 3.1 Reflections
Monitor Your Understanding

Which conversion technique do you like better: set up a proportion or dimensional analysis?

Objective 1 – Compare Measurement and Model Scales
Objective 2 – Understand Unit-less Scales

Objective 1 – Compare Measurement and Model Scales

Key Terms

Scale	
Measurement scale	
Model scale	
Scaling factor	

Summary

Conversions are ratios involving a change in units using measurement scales, keeping the magnitude of the underlying quantity (length, time, etc.) the same. Model scales are proportional relationships between a model and what it actually represents in the world. Model scales indicate a change in magnitude (and also possibly units).

As you read, jot down notes and questions. At the end of this section's Guided Worksheets there is a space for **Reflection** *and* **Monitoring Your Understanding***, where you can try and answer your questions after completing the guided activities and homework.*

Notes & Questions

Guided Practice Activity #1 – Model Scales

1. You wish to create a scale model of the solar system on your campus quad that is circular and measures 240 yards in diameter.

Planet	Orbital Radius (million km)	yards
Mercury	57.9	**1.00**
Venus	108.2	
Earth	149.6	
Mars	227.9	
Asteroids	686.1	
Jupiter	778.3	
Saturn	1429.4	
Uranus	2875	
Neptune	4504.5	
Pluto	5915.8	

a. If you set Mercury to 1 yard, what is the model scale you are using?

b. Use this ratio to determine how far the other planets would be and fill in the table above (2 decimal places of accuracy).

Objective 2 – Understand Unit-less Scales	

Key Terms

Unit-less scale	
Unit-less ratio	

Summary

A **model scale** indicates a change in magnitude between a model and that which it represents (the actual quantity). It is possible for the **units** of both quantities in the **ratio** to be the same, in which case the units cancel and we have a unit-less scale.

Model airplanes are usually on a 1 : 72 scale, meaning 1 inch (or centimeter, miles, etc.) of model length corresponds to 72 inches (or centimeters, miles, etc.) of actual length. This 1:72 scale is increasingly popular world-wide for die-cast toys and Japanese anime characters. In the United States, the 1:48 scale is still more prevalent. Note this scale works for any units (as did the 1:72 scale), 1 centimeter corresponds with 48 centimeters and 1 foot corresponds with 48 feet.

As you read, jot down notes and questions. At the end of this section's Guided Worksheets there is a space for **Reflection** *and* **Monitoring Your Understanding***, where you can try and answer your questions after completing the guided activities and homework.*

Notes & Questions

Guided Practice Activity #1 – Jump Like a Frog

1. In the book, *If You Could Jump Like a Frog*, a 3 inch frog is said to jump 60 inches.

   a. What is the ratio between body length and jump distance, with the first quantity scaled to one?

   b. Is this a unit-less ratio?

c. Assuming jump distance is proportional to body length, find an equation for this relationship.

d. What are the units of the constant of proportionality?

e. Assuming people can jump like frogs, use your equation to determine how far a 4 foot tall child can jump?

f. Assuming people can jump like frogs, use your equation to determine how far a 2 meter tall adult can jump?

2. Now assume we have been given that a 3 inch frog can jump 5 feet.

a. What is the ratio between body length and jump distance (use the 3 and 5 but don't convert units), with the first quantity scaled to one?

b. Is this a unit-less ratio?

c. Assuming jump distance is proportional to body length, find an equation for this relationship.

d. Assuming people can jump like frogs, use your equation to determine how far a 4 foot tall child can jump?

e. Assuming people can jump like frogs, use your equation to determine how far a 2 meter tall adult can jump? Use 1 m = 3.28 ft. (1 decimal place accuracy for answer)

Planet	Orbital Radius (millions km)	miles	millions miles	Earth Set to One			
Mercury	57.9						
Venus	108.2						
Earth	149.6						
Mars	227.9						
Asteroids	686.1						
Jupiter	778.3						
Saturn	1429.4						
Uranus	2875						
Neptune	4504.5						
Pluto	5915.8						

1. First convert the orbital radii into miles using 1 km = 0.621371 miles. What formula could you type into Excel in cell **C2** and fill down assuming Mercury's orbital radii of 57.9 is in cell **B2** (round to whole numbers)?

2. Next convert the miles to millions of miles (2 decimal places). What formula could you type into Excel in cell **D2** and fill down (miles are in Column **C**)?

3. Now convert the orbital radii in millions km to a scale with Earth set to 1 (2 decimal places). What formula could you type into Excel in cell **E2** and fill down?

4. Use Mercury's orbital radii in cell **E2** from the scale in #**3** in a sentence.

5. Now convert the orbital radii in miles and millions miles each to a scale with Earth set to 1. Why are these scales with Earth set to one all the same?

6. Assume you want a model of the solar system with Earth set to 1 foot from the sun. The scale is thus 1 ft : 92,957,102 miles. Using 5,280 ft : 1 mile, what is the unit-less scale for this model of the solar system?

Section 3.2 Reflections
Monitor Your Understanding

Did the *Jump Like a Frog* activity help you to appreciate the power of unit-less scales?

Section 3.3 Rates

Objective 1 – Understand Compound Units

Key Terms

Rate	
Compound units	

Summary

We have just been looking at **unit-less ratios** where both quantities involved had identical **units** which cancelled. If the units are different we often scale the second quantity to a "nice" number (one, one hundred, one thousand, etc.) and create a rate between quantities with different units, with the second quantity scaled to a "meaningful standard" and read using the word "per." Rates will have *compound units* like meters per second (m/s) or miles per gallon (mi/gal), and these require caution when using rates in equations. We must always make sure the units cancel appropriately so that the output has correct units. Dividing by a rate is like multiplying by the reciprocal including the reciprocal of the compound units.

Rates allow for accurate comparisons across time and space. Larger populations would be expected to have more motor vehicle deaths so we use rates to control for different population sizes. A fatality rate per 100,000 people statistic from the U.S. can be compared to one from a much smaller country like Great Britain. Statistics related to societies or countries such as wealth, gross domestic product (GDP), or births are often compared to the population for this reason, giving rates per person or *per capita*.

As you read, jot down notes and questions. At the end of this section's Guided Worksheets there is a space for **Reflection** *and* **Monitoring Your Understanding**, *where you can try and answer your questions after completing the guided activities and homework.*

Notes & Questions

Guided Practice Activity #1 – Prescription Rates

In 2011 nearly 14 million monthly prescriptions were written for American adults 20-39 for ADHD. If there were 82,364 thousand adults aged 20-39 compute the following:

1. The rate per adult aged 20-39 (3 decimal places).

2. If we divide 14 million prescriptions by the rate (prescriptions per adult) from **#1**, what are the units of the answer?

3. The rate of prescriptions per 100 adults 20-39 (1 decimal place).

4. The rate of prescriptions per 1,000 adults 20-39 (1 decimal place).

5. The rate of prescriptions per 100,000 adults 20-39 (1 decimal place).

6. Which rate do you think is "best"?

Guided Practice Activity #2 – Compound Units

For each question determine the answer with proper units. The first one has been done for you.

1. $5100 ÷ 300 $ per month

$$\cancel{\$5100} \times \frac{1 \text{ month}}{\cancel{\$300}} = 17 \text{ months}$$

2. $400,000 \text{ people} \times 6.2m^2 \text{ per person}$

3. 20 miles ÷ 40 miles per hour

4. 450 lbs ÷ 50 lbs per in.^{3}

5. 12.47 MVD per 100,000 drivers × 281,312 thousand drivers (1 decimal place)
    Note: MVD = motor vehicle deaths

6. 1.13 MVD per 100 million VMT × 3095 billion VMT (1 decimal place)
    Note: VMT = vehicle miles traveled

7. $\dfrac{210 \text{ miles}}{30 \text{ mpg}} \times \$2.50/\text{gal}$

Excel Preview – 3.5 Marathons

◢	A	B	C	D	
1		**Marathon Distance:**		26.21875	**miles**
2		miles	plus yards		
3		26	385	42.195	km
4					

	Col H	Col I	Col J	Col K
Hours	**Minutes**	**Minutes/Mile**	**Miles/Minute**	**Miles/Hour**
2.92				
2.88				
2.78				

1. Fill in the table by converting the marathon times from hours into minutes. Then use the distance above to compute the rates (2 decimal places):

   a. Minutes per mile
   b. Miles per minute
   c. Miles per hour

2. The first cell in the table for minutes is **H10**.

   a. What formula can you enter into Excel that could be filled down to convert hours to minutes?

   b. What formula could be entered into cell I10 to compute the rate in minutes per mile?

   c. What formula could be entered into cell J10 to compute the rate in miles per minute?

   d. What formula could be entered into cell K10 to compute the rate in miles per hour?

Section 3.3 Reflections
Monitor Your Understanding

Section 3.4 Dosages and Concentrations

Objective 1 – Compare Dosages and Concentrations
Objective 2 – Understand Density

Objective 1 – Compare Dosages and Concentrations

Key Terms

Concentration	
Dosage	

Summary

An important application of the concepts in this chapter involves medication dosages. Over the counter medications, such as aspirin and cough syrup, come with clearly labeled dosage amounts, such as 2 tablets every two hours or 1 teaspoon every 6 hours for adults 12 years and over. Doctors, nurses, and veterinarians, however, are faced with a vast array of medications in varying concentrations (mg/ml, units/cc, %,…) and dosage amounts (mg/kg, units/lb, g/lb…).

Dilantin at 0.5 g per dose is ordered and is available as 15 mg per 3 ml. Compute how many ml need to be administered. The medication, Dilantin, is a powder that has been mixed into a solution with a concentration of 15 mg per 3 ml. We need to determine how much of the solution contains the 0.5 g we need. First convert 0.5 g to mg:

$$0.5 \text{ g} \cdot \frac{1{,}000 \text{ mg}}{1 \text{ g}} = 500 \text{ mg}$$

Then convert this to ml using the concentration:

$$500 \text{ mg} \cdot \frac{3 \text{ ml}}{15 \text{ mg}} = 100 \text{ ml}$$

Note that we inverted the concentration so that the 15 mg is in the denominator. We could have also set up a proportion here to solve for the ml.

As you read, jot down notes and questions. At the end of this section's Guided Worksheets there is a space for **Reflection** *and* **Monitoring Your Understanding**, *where you can try and answer your questions after completing the guided activities and homework.*

Notes & Questions

Common Conversions

Weight			Volume		
1 gram (g)	=	0.001 kilograms (kg)	1 US fluid ounce	≅	29.57353 milliliters (ml)
1 gram (g)	≅	0.035273962 ounces	1 US cup	=	16 US tablespoons
1 ounce	=	28.34952312 grams (g)	1 US gallon	=	128 US fluid ounces
1 pound (lb)	=	0.45359237 kilograms (kg)	1 liter (l)	≅	33.8140227 US fluid ounces
1 kilogram (kg)	≅	35.273962 ounces	1 milliliter (ml)	=	1 cubic centimeter (cc)
1 kilogram (kg)	≅	2.20462262 pounds (lb)	1 US gallon	=	3.7854 liters

1. Medicine is ordered at 0.048 g and comes mixed in solution at 25 mg/100 ml. How many ml should be administered?

2. A doctor orders 2.3 g of Rocephin to be taken by a 15 lb. 6 oz. infant every 8 hours. The recommended dosage is 75-100 mg/kg per day. Is the ordered amount within the recommended range?

Guided Practice Activity #2 –Rates, Canceling Units, and a Clever Equation

1. Your roommate walks at 3 mph and it takes her 8 minutes to walk to class. You have the same class but walk faster at 5 mph. How long will it take you to get to class?

    a. Convert the 8 minutes into hours (3 decimal places).

    b. How far is class (1 decimal place)?

    c. How long in minutes will it take your roommate to walk to class (1 decimal place)?

    d. Now for the clever equation! Plug both rates (mph) and times (min) into the equation below and confirm that the products are indeed equal.
    $$R_1 * T_1 = R_2 * T_2$$

2. You have a fixed budget for flooring in your new home. One type of flooring, at $4 per square foot, covers 89,280 square inches using your budget. The more expensive flooring at $6 per square foot will cover how much flooring on your budget?
    $$R_1 * A_1 = R_2 * A_2$$

3. Jeannine eats $\frac{1}{4}$ cup oatmeal per day and a container lasts her 36 days. Her husband starts eating $\frac{1}{2}$ cup per day. How long will the container now last with both eating oatmeal each day?
    $$R_1 * T_1 = R_2 * T_2$$

4. A *mole* is a strange unit of measurement giving the amount of a substance that contains 6.022×10^{23} "elementary entities" (atoms, molecules, etc.). The *molarity* is a concentration measured in moles per liter. You have a solution with a molarity of 10^{-3} moles/L and need 100 ml at 10^{-4} moles/L. What do you do?

$$M_1 * V_1 = M_2 * V_2$$

5. You have a solution mixed at a concentration of 5 g/ml and need 100 gal mixed at 4 g/ml. What do you do?

$$C_1 \cdot V_1 = C_2 \cdot V_2$$

6. You wish to count the number of bacteria in 1 ml of a solution but there are too many to count! So you mix the 1 ml with 9 ml of water to dilute it, and then draw 1 ml of the new mixture. Still too many, so you now mix with 9 ml of water and draw 1 ml of this new solution and now count 75 nasty little bacteria swimming around. How many bacteria are in the original solution?

$$C_1 \cdot V_1 = C_2 \cdot V_2$$

Objective 2 – Understand Density

Key Terms

Density	

Summary

The **density** of a material refers to how much mass of the material occupies a fixed space. This explains why certain materials are heavier than others. A block of ice (916.7 kg/m^3) weighs less than a same size block of lead ($11{,}340$ kg/m^3) which in turn weighs less than the same size block of gold ($19{,}300$ kg/m^3). Please refer to the *Densities* sheet in **Data Sets** for the actual densities for these and other materials.

Densities often involve tricky volume units like cubic meters (m^3) and cubic inches (in^3). We have seen that 1 inch = 2.54 cm, so a square inch (in^2), which is a square that measures one inch on a side, will have 2.54 cm on each side. The area of a square inch in cm^2 is thus:

$$1 \text{ in}^2 \;=\; 2.54^2 \text{ cm}^2$$

CAUTION! Many students do not square the 2.54 when converting from square inches to square cm. Technically we square the 1 on the left, but $1^2 = 1$.

Similarly we can compare cubic inches and cubic centimeters:

$$1 \text{ in}^3 \;=\; 2.54^3 \text{ cm}^3$$
$$1 \text{ in}^3 \;=\; 16.4 \text{ cm}^3$$

1 cubic inch = the volume of a cube which is 1 inch on each side
1 cubic cm is much smaller than 1 cubic inch! A cubic inch holds over 16 cubic cm

As you read, jot down notes and questions. At the end of this section's Guided Worksheets there is a space for **Reflection** *and* **Monitoring Your Understanding**, *where you can try and answer your questions after completing the guided activities and homework.*

Notes & Questions

Guided Practice Activity #1 –Densities and Compound Units

Weight			Volume		
1 gram (g)	=	0.001 kilograms (kg)	1 US fluid ounce	$\cong$	29.57353 milliliters (ml)
1 gram (g)	$\cong$	0.035273962 ounces	1 US cup	=	16 US tablespoons
1 ounce	=	28.34952312 grams (g)	1 US gallon	=	128 US fluid ounces
1 pound (lb)	=	0.45359237 kilograms (kg)	1 liter (l)	$\cong$	33.8140227 US fluid ounces
1 kilogram (kg)	$\cong$	35.273962 ounces	1 milliliter (ml)	=	1 cubic centimeter (cc)
1 kilogram (kg)	$\cong$	2.20462262 pounds (lb)	1 US gallon	=	3.7854 liters

1. We do the first one for you demonstrating the dimensional analysis technique. Convert the density of lead to ounces per cubic inch:

$$\frac{11,340 \ \cancel{kg}}{1 \ \cancel{m^2}} \cdot \frac{1 \ \cancel{m^2}}{100^3 \ \cancel{cm^3}} \cdot \frac{2.54^3 \ \cancel{cm^3}}{1 \ in^3} \cdot \frac{35.27 \ oz}{1 \ \cancel{kg}} = 6.55 \ oz/in^3$$

2. Convert the density of water, 1 gram per ml, to pounds per gallon (2 decimal places).

3. Convert the density of gold 19,300 kg/m³ to g/cm³ (1 decimal place).

Excel Preview – 3.2 Densities

	A	B	C	D	E	F
1		lb	kg		cubic in.	cubic m
2		2.2	1		1,000,000	16.387
3						
4				a.)	b.)	c.)
5		Material	Density (kg/m^3)	lb/in.^3	Gold Scale	Water Scale
6		Helium	0.179	6.45E-06	0.001	0.00018
7		Aerographite	0.2	7.21E-06	0.001	0.00020
8		Metallic microlattice	0.9	3.24E-05	0.005	0.00090
9		Aerogel	1	3.61E-05	0.005	0.00100
10		Air	1.2	4.33E-05	0.006	0.00120
11		Tungsten hexafluoride	12.4	4.47E-04	0.064	0.01240
12		Liquid hydrogen	70	2.52E-03	0.362	0.07000
13		Styrofoam	75	2.70E-03	0.388	0.07500
14		Cork	240	8.65E-03	1.242	0.24000
15		Lithium	535	1.93E-02	2.769	0.53500
16		Wood	700	2.52E-02	3.623	0.70000
17		Potassium	860	3.10E-02	4.451	0.86000
18		Sodium	970	3.50E-02	5.021	0.97000
19		Ice	916.7	3.30E-02	4.745	0.91670
20		Water (fresh)	1,000	3.61E-02	5.176	1.00000
21		Water (salt)	1,030	3.71E-02	5.331	1.03000

1. To convert the given densities from kg/m^3 to lb/in.^3 what formula is entered into cell **D6** and filled down?

2. The density of water is given in row **20**. In column **F** we wish to scale the density of water to 1 and all other densities proportionally. What formula is entered in cell **F6** and filled down?

3. What are the units of the densities in column **F**?

4. The following column chart shows the densities in kg/m^3. If we graphed the water scale column F and put it side by side with this column chart would both charts look the same?

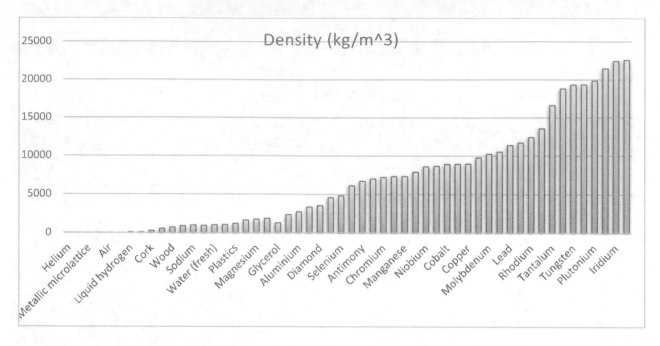

5. Instead of 25,000 what would be the maximum value on the vertical axis for the Water Scale?

Section 3.4 Reflections
Monitor Your Understanding

Were you surprised by the challenging nature of determining correct dosages and concentrations?

How important is quantitative literacy for people who want a career as a nurse or veterinarian?

Objective 1 – Understand Standard Error of the Mean

Key Terms

Distribution of sample means	
Standard error of the mean	
Three essences of variability	
Two types of variability	

Summary

The following table summarizes height data for women:

Table 10. Height in inches for females aged 20 and over and number of examined persons, mean, standard error of the mean, and selected percentiles, by race and Hispanic origin and age: United States, 2011–2014

Race and Hispanic origin and age	Number of examined persons	Mean	Standard error of the mean	Percentile								
				5th	10th	15th	25th	50th	75th	85th	90th	95th
All racial and Hispanic-origin groups[1]							Inches					
20 years and over	5,547	63.7	0.08	59.0	60.1	60.7	61.7	63.7	65.5	66.5	67.2	68.3
20–29 years	928	64.1	0.12	59.8	60.8	61.4	62.3	64.1	65.8	66.6	67.4	68.6
30–39 years	957	64.3	0.12	59.5	60.5	61.3	62.3	64.4	66.4	67.3	67.9	69.0
40–49 years	987	64.1	0.12	59.4	60.5	61.2	62.2	64.2	65.9	66.8	67.6	68.7
50–59 years	924	63.7	0.15	59.4	60.2	60.9	61.8	63.9	65.3	66.3	67.1	68.2
60–69 years	888	63.2	0.15	59.0	60.1	60.5	61.5	63.2	64.9	65.8	66.5	67.7
70–79 years	527	62.7	0.14	58.1	59.2	59.8	60.8	63.0	64.5	65.4	66.1	67.1
80 years and over	336	61.3	0.15	56.9	57.7	58.3	59.6	61.3	63.1	63.7	64.3	65.5

We were given the sample mean height of 63.7 inches from a sample of 5547 women, but instead of giving the standard deviation they give the statistic called the standard error, 0.08. If we took another sample of women we would get a different sample mean. The variability in these sample means is measured by the standard error.

The distribution of sample means refers to the dataset we could create by repeatedly sampling 5547 random representative women and computing their mean height. Every time we do this we would get a different sample mean height. This is an important and subtle point that deserves special care! Recall the distributions of flight delays and average flight delays by airport.

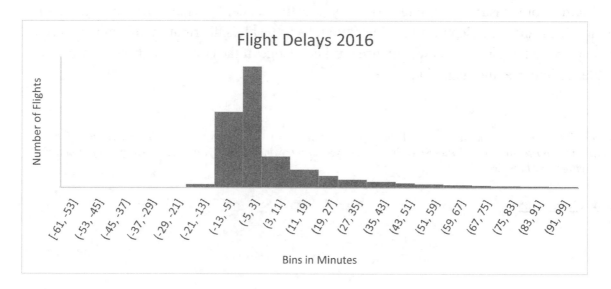

The first histogram is showing the distribution for over 1 million flights with delay less than 100 minutes and it is clearly right skewed, while the second histogram is showing the distribution of the 288 average airport delays. The distribution of averages is approximately normal, and this hints at the Central Limit Theorem (discussed in Chapter 12) that tells us the distributions of sample means will be normal (assuming samples are of size 30 or more) no matter what the shape of the original distribution! In addition the distribution of averages will be much less spread out ($\sigma = 7.81$ for airports) than the original distribution ($\sigma = 38.95$ for flights).

Statistics is the study of variability. We characterize variability by describing the shape of the distribution (bell shape, skewed, etc.), the center of the distribution (mean, median, mode), and the spread of the distribution (standard deviation, range, count, max, min). The shape, center, and spread are the three essences of variability. There are two types of variability we encounter: natural variability (heights, IQ-scores etc.) and sampling variability (e.g. different means from repeated sampling).

The standard error measures sampling variability; it is the standard deviation of the distribution of the sample means. The Central Limit Theorem tells us that this distribution of sample means will be approximately normal, so we can use our knowledge of the normal distribution and percentiles to interpret the standard error.

As you read, jot down notes and questions. At the end of this section's Guided Worksheets there is a space for **Reflection** *and* **Monitoring Your Understanding**, *where you can try and answer your questions after completing the guided activities and homework.*

Notes & Questions

Guided Practice Activity #1 – z-scores and Standard Error

The histograms we saw above are discrete versions of idealized continuous distributions, such as the following normal distribution:

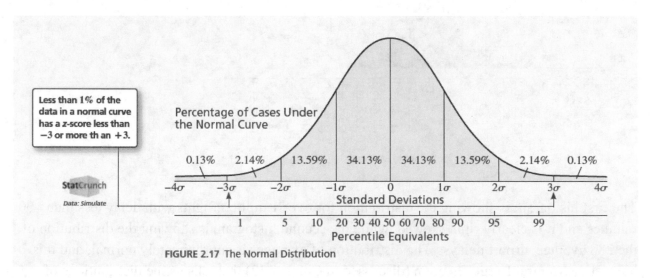

FIGURE 2.17 The Normal Distribution

1. In the normal distribution what percentage of data values fall below (to the left of) the mean, μ?

   Below $+1\sigma$ (= +1 standard deviation)?

   Below $+2\sigma$?

   Below $+3\sigma$?

2. On the normal curve above what percentage of data values lie within two standard deviations of the mean (from -2σ to $+2\sigma$)?

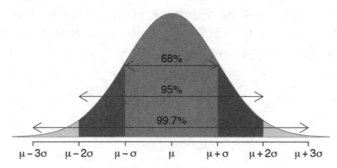

3. Here are some fun questions. Make an educated guess ☺:

   a. 95% of US women fall between what two heights?

   b. 95% of US men fall between what two heights?

4. Use the following table for women to answer the questions:

Table 10. Height in inches for females aged 20 and over and number of examined persons, mean, standard error of the mean, and selected percentiles, by race and Hispanic origin and age: United States, 2011–2014

Race and Hispanic origin and age	Number of examined persons	Mean	Standard error of the mean	5th	10th	15th	25th	50th	75th	85th	90th	95th
All racial and Hispanic-origin groups[1]							Inches					
20 years and over	5,547	63.7	0.08	59.0	60.1	60.7	61.7	63.7	65.5	66.5	67.2	68.3
20–29 years	928	64.1	0.12	59.8	60.8	61.4	62.3	64.1	65.8	66.6	67.4	68.6
30–39 years	957	64.3	0.12	59.5	60.5	61.3	62.3	64.4	66.4	67.3	67.9	69.0
40–49 years	987	64.1	0.12	59.4	60.5	61.2	62.2	64.2	65.9	66.8	67.6	68.7
50–59 years	924	63.7	0.15	59.4	60.2	60.9	61.8	63.9	65.3	66.3	67.1	68.2
60–69 years	888	63.2	0.15	59.0	60.1	60.5	61.5	63.2	64.9	65.8	66.5	67.7
70–79 years	527	62.7	0.14	58.1	59.2	59.8	60.8	63.0	64.5	65.4	66.1	67.1
80 years and over	336	61.3	0.15	56.9	57.7	58.3	59.6	61.3	63.1	63.7	64.3	65.5

a. What is the probability that a random woman 20 years and over is over 66.5 inches tall.

b. What is the standard deviation of women's heights 20 years and over? Recall that the 85th percentile is roughly 1 standard deviation above the mean.

c. Use the standard deviation to determine the 95% interval about the sample mean for women's heights 20 years and over. How close was your guess from #3?

d. The distribution of sample means ($n = 5547$) will be centered at the true population mean, μ, but will have a standard deviation given by the standard error, 0.08 inches. There is a 95% chance our sample mean of 63.7 inches will be within two standard errors of the mean, μ. Use the standard error to determine the 95% interval about the true population mean for women's heights 20 years and over (2 decimal places).

e. Recall we saw above that there is a 15% chance a woman will have a height of 66.5 inches or more. Compute the z-score for a sample mean of 66.5 inches assuming $\mu = 63.7$. What is the probability that a random sample ($n = 5547$) of 20-29 year old women has a mean height over 66.5 inches?

5. Interpret the standard error for men 20 years and over in a sentence using 95% probability.

Table 12. Height in inches for males aged 20 and over and number of examined persons, mean, standard error of the mean, and selected percentiles, by race and Hispanic origin and age: United States, 2011–2014

Race and Hispanic origin and age	Number of examined persons	Mean	Standard error of the mean	5th	10th	15th	25th	50th	75th	85th	90th	95th
All racial and Hispanic-origin groups[1]							Inches					
20 years and over	5,232	69.2	0.08	64.3	65.4	66.1	67.2	69.1	71.2	72.3	73.0	74.1
20–29 years	937	69.4	0.10	64.9	65.7	66.3	67.4	69.4	71.4	72.6	73.4	74.3
30–39 years	914	69.5	0.12	64.5	65.8	66.5	67.5	69.5	71.5	72.6	73.5	74.4
40–49 years	872	69.4	0.17	64.5	65.7	66.3	67.3	69.3	71.2	72.6	73.1	73.9
50–59 years	852	69.3	0.20	64.8	65.7	66.3	67.2	69.1	71.4	72.3	73.0	74.2
60–69 years	877	69.0	0.18	63.8	65.2	65.9	67.1	69.1	71.2	72.1	72.5	73.8
70–79 years	486	68.1	0.12	63.9	64.7	65.1	66.3	67.9	69.8	70.9	71.7	72.9
80 years and over	294	67.6	0.23	62.9	64.2	64.7	65.9	67.7	69.4	70.5	71.0	72.0

6. Determine the z-score for your height.

7. Who is taller: a 6 foot tall woman or a 6 foot 6 inch tall man?

The Black Swan by Nassim Taleb

Why does he call the Normal Distribution the Great Intellectual Fraud?

Because socio-economic data is usually not normal but economists' models all use the normal distribution!

Example: US Household Income 2005

$\mu = \$84,800$ and $\sigma = \$386,000$ Note: median value = $58,500

1. What does the mean being greater than the median tell you about the data set?

2. The top 0.01% of households averaged $35,473,200. Compute the corresponding z-score (1 decimal place).

3. U.S. Women 20 years and older have an average height of 63.7 inches with a standard deviation of 2.8 inches. How tall would a woman be if she had the same z-score computed in #2 above (1 decimal place)?

4. What is a "black swan" a metaphor for in economic terms?

Excel Activity – Uniform Distributions and Random Sampling

Instead of previewing an Excel problem, we give an activity that explores the distribution of sample means and the Central Limit Theorem (covered in detail in chapter 12).

1. In this activity you will be creating a data set that consists of 9,000 random digits from 1 – 9. Sketch what you think the **histogram** will look like for this large data set. Be sure to label your bins and also the scale on the vertical axis.

2. What do you think is the **mean** and **standard deviation** of your data set?

3. Open Excel and in **Column A** enter the series of numbers from 1 – 9000. To do this without filling with the mouse keypad use the **Fill** shortcut in the **Editing** menu.

   a. First enter 1 in cell **A1** and make sure **A1** is highlighted,
   b. then choose the **Series** option.
   c. Then choose **Columns** and set the **Step** and **Stop** Values. Click **OK**!

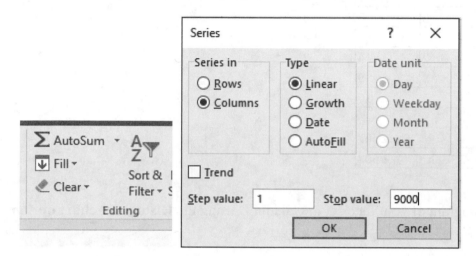

4. In **Column B** we will enter random digits from 1 – 9 using **=RANDBETWEEN(1,9).** Do so and fill down by double clicking on the fill handle. Compute the mean and standard deviation using **AVERAGE** and **STDEV.P**

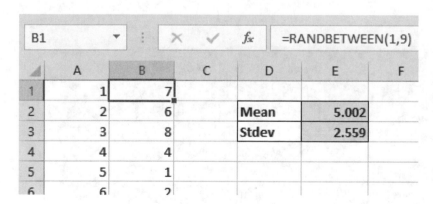

5. Compute the actual standard deviation by computing the average squared deviation from the 9 possible data values and then use **SQRT** to find the standard deviation. Recall the deviation is computed by subtracting the mean. You should create space in your spreadsheet that looks like this:

data	deviation	dev^2
1	-4	16
2	-3	9
3	-2	4
4	-1	1
5	0	0
6	1	1
7	2	4
8	3	9
9	4	16
	SUM/9:	6.666667
	STDEV	2.582

6. Create a histogram of your data set in Column B using the **Histogram chart option**. Is this what you sketched in #1?

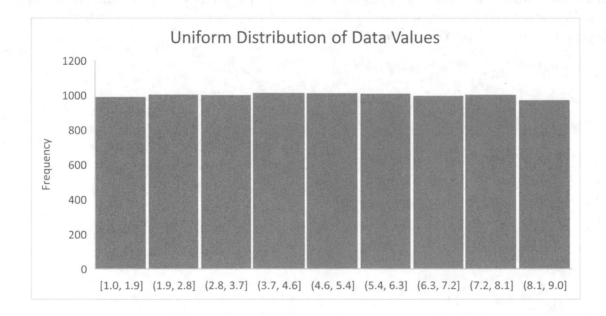

7. Now we will create the **distribution of sample means**. We will make **1,000 samples** of size **30** and compute each sample mean. Open a new spreadsheet and make it look like this:

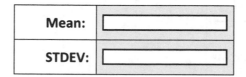

	A	B	C	D	E	F	G	H	I	J	K	L	M	N	O	P	Q	R	S	T	U	V	W	X	Y	Z	AA	AB	AC	AD	AE	AF
	1	1	2	3	4	5	6	7	8	9	10	11	12	13	14	15	16	17	18	19	20	21	22	23	24	25	26	27	28	29	30	MEAN
2	2	8	8	6	7	8	1	6	1	3	9	7	2	7	3	9	2	8	7	7	7	7	4	6	1	8	5	2	8	2	9	5.60
3	3	4	9	6	3	6	1	2	8	7	8	7	6	6	9	3	4	6	6	9	6	2	6	1	7	8	1	5	9	1	9	5.50
4	4	7	6	3	8	1	7	8	6	8	1	4	3	1	7	2	6	8	4	9	6	2	3	7	4	7	9	1	4	3	6	5.03
5	5	8	5	6	6	1	9	5	9	9	4	9	5	8	1	8	8	4	9	6	3	7	3	4	9	1	4	3	5	9	9	5.90
6	6	8	5	6	8	8	2	8	6	2	4	2	4	3	1	7	6	3	4	4	7	3	5	5	5	4	5	1	2	1	4	4.43

8. Now compute the **mean** and **standard deviation** of all 1,000 of the sample means in **Column AF**. How do these compare to the mean of your data set?

Mean:	
STDEV:	

9. What do think the histogram of the distribution of the sample means will look like? The same as the distribution of data set?

10. Create a histogram of the sample means using the **Histogram chart option**. What kind of distribution is this? Are you surprised?!

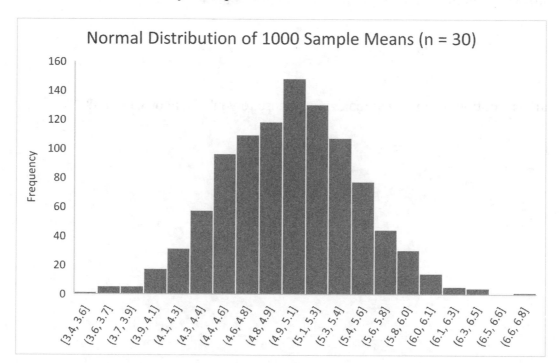

11. The Central Limit Theorem tells us the distribution of sample means will be normally distributed as long as the original distribution is approximately normal or the sample size 30 or more. The distribution of sample means will have the same center but will be less spread out:

$$\mu_{\bar{X}} = \mu \qquad \sigma_{\bar{X}} = \frac{\sigma}{\sqrt{n}}$$

Compute the standard deviation of the distribution of sample means by dividing the data set standard deviation from #5 by the square root of the sample size.

Mean:		
STDEV:		
Sigma/SQRT(n):		

Section 3.5 Reflections
Monitor Your Understanding

What was surprising about the notion of the distribution of sample means?

Is the normal distribution beginning to make more sense to you? Why or why not?

Chapter 4 – Percentages

Section 4.1 Percentages

Objective 1 – Understand Rates per 100
Objective 2 – Represent Percentages as Fractions and Decimals

Objective 1 – Understand Rates per 100

Key Terms

Percentage	
Rate of return	
Annuity	

Summary

Percentages are perhaps the most common type of **ratio**, and we have already seen several examples including the **APR** associated with loans in Chapter 1. In Chapter 3 we studied **rates**, and percentages are a special type of rate with the second quantity scaled to 100. Note that rates of return (and interest rates) are an exception to the rule that rates are ratios comparing different **units**. Taking 3% of a quantity means we multiply the quantity by 3%.

As you read, jot down notes and questions. At the end of this section's Guided Worksheets there is a space for **Reflection** *and* **Monitoring Your Understanding**, *where you can try and answer your questions after completing the guided activities and homework.*

Notes & Questions

Guided Practice Activity #1 – Rates of Return

The following graphic gives the rates of return for the period **ending** 3/31/2018.

GUARANTEED
TIAA Traditional Annuity - Group Retirement Annuity
Rates of Return for the Period Ending 3/31/2018

Current Rates	1 Year	3 Year	5 Year
4.00%	∧ 3.75%	∧ 3.92%	∧ 3.98%
AS OF CLOSE 3/31/2018			

The TIAA Traditional Annuity is a guaranteed annuity account backed by TIAA's claims-paying ability. It guarantees your principal and a contractually specified minimum interest rate, plus it offers the opportunity for additional amounts in excess of this guaranteed rate. These additional amounts are declared on a year-by-year basis by the TIAA Board of Trustees.

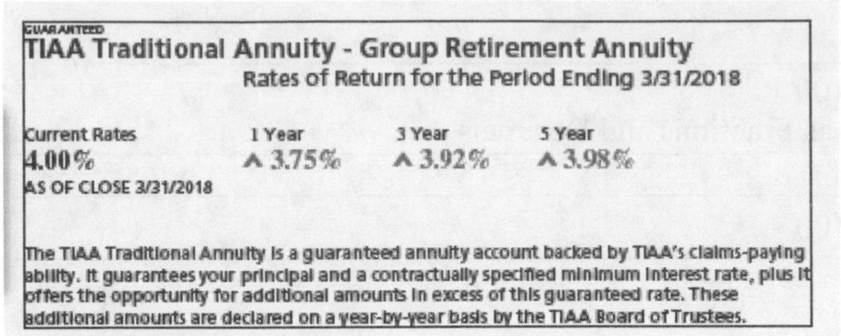

FIGURE 4.1 TIAA Traditional Annuity Rates of Return
Data from: Values from TIAA Traditional Annuity. Published by TIAA-CREF, © 2018.

1. Assume someone had $100,000 in their retirement account on 3/3/1/**2017**.

   a. Use the 1 year rate to calculate how much interest they gain in this 1 year period until 3/31/2108.

   b. Use the 1 year rate to calculate how much money they had on 3/31/2018.

2. Assume someone had $100,000 in their retirement account on 3/3/1/**2013**. The **five year** rate of 3.98% means that each year their initial balance will grow by 3.98%. To find out how much money they will have in their account five years later we can fill in the following table:

	Five Year Rate 3.98%		
Date	Initial Principal	Interest	Ending Principal
3/31/2013			
3/31/2014			
3/31/2015			
3/31/2016			
3/31/2017			
3/31/2018			

Summary

The fact that a **percentage** is a **rate** per 100 allows us to write percentages as fractions and hence decimals. This leads to memorizing rules about moving the decimal point two places left or right, which can be easy to apply for percentages such as $75\% = 75/100 = 0.75$; but may cause confusion when working with percentages less than 1% ($0.75\% = ??$) or greater than 100% ($750\% = ??$). We will always focus on the rate per 100 aspect to help avoid this confusion.

CAUTION! Many people incorrectly state that to write a decimal as a percentage *multiply by 100*. The correct statement is to *multiply by 100%* which equals one. This moves the decimal point two places right and inserts the % symbol.

As you read, jot down notes and questions. At the end of this section's Guided Worksheets there is a space for **Reflection** *and* **Monitoring Your Understanding***, where you can try and answer your questions after completing the guided activities and homework.*

Notes & Questions

Guided Practice Activity #1 – Move the Decimal

1. Fill in the following table. Note that the numerals in the first column are the numerators in the fractions. Dividing by 100 moves the decimal point two places to the left in the third column.

Percentages	Fractions	Decimals
75%	$\dfrac{75}{100}$	0.75
	$\dfrac{0.75}{100}$	
		7.5
4.5%		
	$\dfrac{275}{100}$	
		−0.0023
100%		
		3

2. We typically encounter percentages less than one when dealing with **periodic rates** for loans. For each of the following compute the periodic interest rate (%) and the periodic interest ($) for the first period. Note: avoid round-off error by using full decimal expansion (Excel is great for this!).

   a. APR = 8%, monthly, Balance = $2,500

   b. APR = 20%, monthly, Balance = $2,500

c. APR = 18%, daily, Balance = $1,000

d. APR = 14.5%, daily, Balance = $12,000

e. APR = 22.3%, annually, Balance = $4,892.33

Excel Preview – 4.2 Budget

The Fiscal Year (FY) projected 2019 budget is given in the spreadsheet broken down by Outlays (Mandatory and Discretionary spending) and Receipts (income).
https://www.whitehouse.gov/omb/historical-tables/

	A	B	C	D	E	F	G	H	I
1	Fiscal Year (FY) 2019 Projected Budget (billions $)								
2									
3									
4	Outlays (Expenditures):					% of Sub-total		% of Total	
5	Mandatory					b.)	b.)	b.)	b.)
6	Social Security	1046.5							
7	Medicare	624.7							
8	Medicaid	412.0							
9	Other	656.0		a.)					
10			Sub-total:						
11	Discretionary					b.)	b.)	b.)	b.)
12	National Defense	678.2							
13	Non-Defense	626.0		a.)					
14			Sub-total:					b.)	b.)
15	Interest	363.4		a.)					
16			Sub-total:			TOTAL:	a.)		
17									
18	Receipts (Income):					b.)	b.)	b.)	b.)
19	Individual Income Tax	1,687.7							
20	Corporate Income Tax	225.3							
21	Payroll Tax	1,237.7							
22	Excise Tax + Other	271.6		a.)					
23			Sub-total:						
24									
25						TOTAL:	a.)		
26									

1. How much money was projected to be spent on social security in 2019?

2. Assuming a population of 330 million people how much is this per capita?

3. Assuming 15% of population is over 65, how much is this per elderly person (65+)?

4. Payroll taxes are supposed to fund the bulk of social security, Medicare and Medicaid. In 2019 payroll taxes are what percentage of these three outlays?

5. Compute the sub-total for the mandatory outlays and % of this sub-total for each mandatory outlay as a decimal to 3 places and a percentage to 1 place.

Outlays (Expenditures):		% of Sub-total	
Mandatory		**b.)**	**b.)**
Social Security	1046.5		
Medicare	624.7		
Medicaid	412.0		
Other	656.0		
Sub-total:	**a.)**		

6. What formula can you enter in cell **F6** (see spreadsheet or prior page) that can be filled down?

7. How can you represent a decimal as a percentage in Excel without multiplying by 100?

Section 4.1 Reflections
Monitor Your Understanding

What have you learned about percentages that was new or different?

Section 4.2 Part-to-Part Ratios

Objective 1 – Understand Part-to-Part Ratios
Objective 2 – Evaluate Concentrations

Objective 1 – Understand Part-to-Part Ratios

Summary

The most common interpretation of a **percentage** is as a **part-to-whole ratio**. Indeed many people define percentages as part-to-whole ratios, which can cause much confusion when dealing with other uses of percentages. If we are given the ratio of boys to girls in a class is 2 : 3, then it can be easy to mistakenly say: "The boys are 66.7% of the class.", or even worse: "66.7% of the girls are boys!" The correct statement uses the size of each part: "The number of boys is 66.7% of the number of girls."

Boys (part)	*Girls (part)*	*Students (whole)*
2	3	5
66.7	100	
40	**60**	100

We can add the parts to get the whole, 2 boys : 3 girls : 5 students; and then scaling the whole to 100, we get 40 boys : 60 girls : 100 students. So the boys are 40% of the class not 66.7%. To summarize we use each of the numbers in the first column in a sentence.

- There are 2 boys for every 3 girls in the class.
- There are 66.7 boys for every 100 girls, so the number of boys is 66.7% of the number of girls.
- There are 40 boys for every 100 students, the boys are 40% of the class.

As you read, jot down notes and questions. At the end of this section's Guided Worksheets there is a space for **Reflection** *and* **Monitoring Your Understanding**, *where you can try and answer your questions after completing the guided activities and homework.*

Notes & Questions

Guided Practice Activity #1 – Parts of the Pop

The following table gives sizes of the U.S. population on July 1, 2017 (https://www.census.gov/data/datasets/2017/demo/popest/nation-detail.html).

Under 18 years	18 to 64 years	65 years and over	Total
73,655,378	201,205,121	50,858,679	[ ]
[ ]	100.0	[ ]	
[ ]	[ ]	100.0	
[ ]	[ ]	[ ]	100

1. Compute the total, and then fill in the rest of the ratio table given the different quantities that have been set to 100. Make sure to always use the numbers in the first row to avoid round-off error. Note the header row with titles is not counted as the "first" row.

2. Use each of the 4 numbers from the first column in a sentence. One sentence for each number, and interpret the as both rates per 100 and as percentages for the three rates.

Key Terms

Dilution	

Summary

A solution with a concentration of 0.5% means there is 0.5 parts of medicine for every 100 parts of solution. Dilution is an important concept for anyone interested in the biological sciences. In general, if we add 100 ml of saline water to 100 ml of a 0.5% sedative solution, we will have 0.5 ml sedative to 200 ml solution, reducing the concentration to 0.25%:

$$\frac{0.5}{200} = \frac{0.25}{100} = 0.25\%$$

As you read, jot down notes and questions. At the end of this section's Guided Worksheets there is a space for **Reflection** *and* **Monitoring Your Understanding**, *where you can try and answer your questions after completing the guided activities and homework.*

Notes & Questions

Guided Practice Activity #1 – Sedative

Your veterinarian is administering a sedative to your 40 pound dog Bella. The sedative is mixed in saline solution. Unfortunately the solution is pre-mixed at 2.5% for a horse and it needs to be a 0.5% concentration to be administered at 10 ml per pound for your dog. How much saline water needs to be added to the current solution to reduce the concentration to 0.5% and have the correct dosage for Bella?

Part	Whole #1 (2.5%)	Whole #2 (0.5%)	
Sedative (ml)	Original Solution (ml)	New Solution (ml)	Dog (lb)
#1. [＿＿]	100		
#2. [＿＿]		100	
		#3. [＿＿]	1
#5. [＿＿]	#6. [＿＿]	#4. [＿＿]	40

1. Put the given information for the original solution by filling in the number of ml associated to 100 ml of original solution.

2. Put the given information for the new solution by filling in the number of ml associated to 100 ml of new solution.

3. Put in the number of ml to be administered per pound of Bella's weight.

4. Determine how many ml of new solution we will need to administer to Bella.

5. Next determine how much sedative will be in the amount from **#4**.

6. Determine how many ml of original solution will contain the amount of sedative from **#5**.

7. Finally determine how much saline water needs to be added to the current solution to reduce the concentration to 0.5% and have the correct dosage for Bella.

Excel Preview – 4.9 Earth

In this excel problem you are given the areas in km^2 of the continents and oceans.

	A	B	C	D
1		1 km =	0.621371 miles	
2			a.)	a.)
3	Continent	Total Area (square km)	Total Area (square miles)	Percentage of Land
4	Asia	43,820,000		
5	Europe	10,180,000		
6	Africa	30,370,000		
7	North America	24,490,000		
8	South America	17,840,000		
9	Australia	9,008,500		
10	Antarctica	13,720,000		
11	Total Land:			a.)
12				
13			b.)	b.)
14	Oceans	Total Area (square km)	Total Area (square miles)	Percentage of Ocean
15	Pacific	155,557,000		
16	Atlantic	76,762,000		
17	Indian	68,556,000		
18	Southern	20,327,000		
19	Arctic	14,056,000		
20	Total Ocean:			b.)
21				
22	Earth:			c.)

1. What formula is entered in cell **C4** that can be filled down?

2. What formula is entered in cell **B11** that can be filled right?

3. What formula is entered in cell **D4** that can be filled down?

4. Does it matter if you use column **B** or **C** to compute the percentage? Why or why not?

5. In the pie chart identify the wedge for the Pacific and Asia, and compute their percentage of the Earth (1 decimal place).

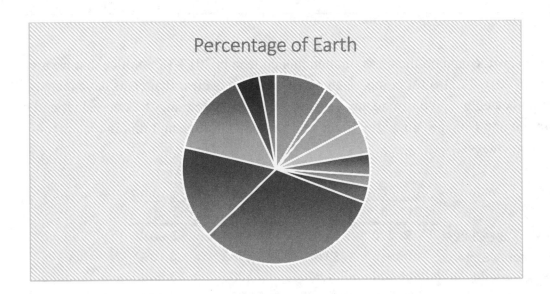

Section 4.2 Reflections
Monitor Your Understanding

Are you surprised by anything in this section?

Section 4.3 Percent Change And Difference

Objective 1 – Compare Total and Relative Change	
Objective 2 – Compare Total and Relative Difference	
Objective 3 – Use Growth/Decay Rates/Factors	

Objective 1 – Compare Total and Relative Change

Key Terms

Total change	
Percent change	

Summary

In Chapter 2 we saw the data on the costs of attending college in 2017-18. Notice that we are given the costs in 2016-17 also for comparison purposes and it is clear that costs have increased in each sector and category. We can quantify this increase by subtracting the two values and get the **Total Change** or $ Change. In addition, the table lists **% Change** and this is our next application of percentages.

	Tuition and Fees			
	2017-18	2016-17	$ Change	% Change
Public Two-Year In-District	$3,570	$3,470	$100	2.90%
Public Four-Year In-State	$9,970	$9,670	$300	3.10%
Public Four-Year Out-of-State	$25,620	$24,820	$800	3.20%
Private Nonprofit Four-Year	$34,740	$33,520	$1,220	3.60%

Many find percent change confusing and ask why bother with it if we have total change. Look again at the data table. Notice that the tuition and fees for *Public Four-Year Out-of-State* undergraduate students increased by $800 compared to only a $100 increase for *Public Two-Year In-State*; but the % Change for *Four-Year* is about the same, 3.2% compared to 2.9% for the *Two-Year*. The percent change puts the total change in perspective by comparing it to the original value, once again demonstrating the power of ratios. An increase of $800 is significant if we start with $1,000, but not so much if we start with $100,000.

Thus tuition and fees have increased by 3.2% from 2016-17 to 2017-18 for *Public Four-Year Out-of-State* undergraduate students; or the 2017-18 tuition and fees are 3.2% more than the 2016-17 tuition and fees. An arrow diagram is a nice way to keep track of the quantities involved:

$$+\$800$$

\$24,820 $\longrightarrow$ **\$25,620**

2016–17 (Original)   **+3.2%**   2017–18 (New)

As you read, jot down notes and questions. At the end of this section's Guided Worksheets there is a space for **Reflection** *and* **Monitoring Your Understanding***, where you can try and answer your questions after completing the guided activities and homework.*

Notes & Questions

Guided Practice Activity #1 – Income

The Census Bureau Report[1] *Income and Poverty in the U.S. 2017* has the following table showing income in 2017 dollars for households:

---

[1] https://www.census.gov/content/dam/Census/library/publications/2018/demo/p60-263.pdf

Characteristic	2016			2017		
		Median income (dollars)			Median income (dollars)	
	Number (thou-sands)	Estimate	Margin of error[1] (±)	Number (thou-sands)	Estimate	Margin of error[1] (±)
HOUSEHOLDS						
All households	126,224	60,309	733	127,586	61,372	552
Type of Household						
Family households....................	82,827	76,676	707	83,088	77,713	836
Married-couple......................	60,804	88,929	710	61,241	90,386	820
Female householder, no husband						
present............................	15,572	41,909	890	15,423	41,703	746
Male householder, no wife present.....	6,452	59,299	2,219	6,424	60,843	1,733
Nonfamily households	43,396	36,530	477	44,498	36,650	557
Female householder	22,858	31,230	616	23,481	30,748	633
Male householder....................	20,539	42,647	716	21,017	44,250	2,186
Race[2] and Hispanic Origin of Householder						
White	99,400	63,188	561	100,065	65,273	685
White, not Hispanic..................	84,387	66,440	857	84,681	68,145	1,050
Black................................	16,733	40,340	1,212	16,997	40,258	949
Asian	6,392	83,183	1,958	6,735	81,331	1,962
Hispanic (any race)...................	16,915	48,700	1,137	17,318	50,486	721
Age of Householder						
Under 65 years.......................	94,425	67,917	593	94,613	69,628	917
15 to 24 years......................	6,238	42,551	1,170	6,211	40,093	1,430
25 to 34 years......................	20,109	62,243	819	20,264	62,294	1,051
35 to 44 years......................	21,500	76,082	1,873	21,576	78,368	1,578
45 to 54 years......................	22,808	78,874	1,181	22,542	80,671	1,064
55 to 64 years......................	23,770	66,642	1,337	24,020	68,567	1,587
65 years and older..................	31,799	40,679	928	32,973	41,125	839

1. Compute the total and percent change for number of all households from 2016 to 2017 by drawing the arrow diagram (numbers in thousands).

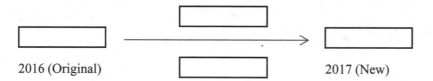

2016 (Original)   2017 (New)

2. Compute the total and percent change for median income for all households from 2016 to 2017 by drawing the arrow diagram.

3. Which of the racial/Hispanic groups had the largest percent change (positive or negative in absolute terms) in median income in 2017?

Objective 2 – Compare Total and Relative Difference	

Key Terms

Total difference	
Percent difference	

Summary

We compute the percent change when we are referring to the same quantity changing over time. If we are comparing the size of two different quantities we compute the total and percent difference.

As you read, jot down notes and questions. At the end of this section's Guided Worksheets there is a space for **Reflection** *and* **Monitoring Your Understanding***, where you can try and answer your questions after completing the guided activities and homework.*

Notes & Questions

Guided Practice Activity #1 – Income Dos

Use the Census Bureau Report[2] *Income and Poverty in the U.S. 2017* table from the previous guided activity showing income in 2017 dollars for households.

1. Compute the total and percent difference in the number of family and non-family households in 2017 (numbers in thousands). Do this in two ways switching the order in the arrow diagram.

Family   Nonfamily

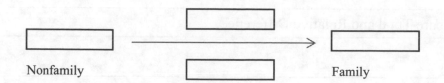

Nonfamily → Family

2. Use the both of percent changes from #1 in sentences.

3. Compute the total and percent difference in the median income for households under 65 and 65 plus in 2017. Do this so you can say: "_____ is ____% more than ____."

65 + → Under 65

Key Terms

Growth/decay rate	
Growth/decay factor	

Summary

The Dallas-Fort Worth Metropolitan area had an estimated population of 5,161,544 people according to the 2000 Census, and it grew by 23.4% over the ensuing decade. What was the estimated population in 2010? We can draw an arrow diagram with x representing the unknown new population in 2010:

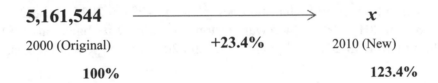

$$\begin{array}{ccc} \mathbf{5,161,544} & \xrightarrow{\hspace{3cm}} & x \\ \text{2000 (Original)} & +23.4\% & \text{2010 (New)} \\ \mathbf{100\%} & & \mathbf{123.4\%} \end{array}$$

It is helpful to represent the original value with 100%. If this grows by 23.4% then the new value will be 123.4%. We are now ready to solve for the new value using two different solution techniques:

Solution #1: The new value is the original value plus 23.4% of the original:

$$x = \mathbf{5,161,544} + 23.4\% \cdot \mathbf{5,161,544}$$

Factoring out the 5,161,544 we get a very important simplification:

$$x = \mathbf{5,161,544} \cdot (1 + 23.4\%)$$
$$x = 5,161,544 \cdot (\mathbf{1.234})$$
$$= 6,369,345$$

The **growth rate** of 23.4% has a **growth factor** of 1.234. If something grows by 23.4%, we multiply it by 1.234.

Solution #2: The arrow diagram above looks like a ratio table, and we can use the values to set up a proportion:

$$\frac{5,161,544}{100} = \frac{x}{123.4} \rightarrow x = \frac{123.4}{100} \cdot 5,161,544 = 6,369,345$$

As you read, jot down notes and questions. At the end of this section's Guided Worksheets there is a space for **Reflection** *and* **Monitoring Your Understanding***, where you can try and answer your questions after completing the guided activities and homework.*

Notes & Questions

Guided Practice Activity #1 – Income Tres

Use the Census Bureau Report[3] *Income and Poverty in the U.S. 2017* table from the previous guided activity showing income in 2017 dollars for households.

1. One reason for concern in this report is that incomes grew by 3.2% in 2016 and 5.2% in 2015, but only by 1.8% in 2017. Work backwards (don't literally write backwards :O) to determine the median income in 2017 dollars in 2015 and 2014 using growth factors.

???	$\longrightarrow$	**$60,309**
2015 (Original)	**+3.2%**	2016 (New)

???	$\longrightarrow$	**???**
2014 (Original)	**+5.2%**	2015 (New)

2. Answer question #1 but now use the proportion technique by setting the original value to 100%.

---

[3] https://www.census.gov/content/dam/Census/library/publications/2018/demo/p60-263.pdf

3. Finally let's consider poverty levels. In 2017 about 12.3% of Americans or 39.7 million people lived below the poverty line.

   a. Compute how many Americans there were using this information.

   b. The number of people living in poverty in 2017 was 2.3% less than the number in 2016. What is the decay factor associated to this decay rate of -2.3%?

   c. Use the decay factor to determine the number in poverty in 2016.

Excel Preview – 4.4 Urban

In this excel problem you are given the U.S. total population and the percentages living in urban and rural areas.

	A	B	C	D	E	F	G	H
1	Census	Total		Increase	Urban	Rural		
2	Year	Population	Increase	%	%	%	Urban	Rural
3	1790	3,929,214	-	-	5.1	94.9		
4	1800	5,308,483			6.1	93.9		
5	1810	7,239,881			7.3	92.7		
6	1820	9,638,453			7.2	92.8		
7	1830	12,860,702			8.8	91.2		
8	1840	17,063,353			10.8	89.2		

1. What formula would you enter in cell **C4** to compute the total change in population?

2. What formula would you enter in cell **D4** to compute the percent change in population?

3. Note that columns **E & F** are labeled %, so the 5.1 in cell **E3** is literally the number 5.1, not 0.051 formatted as a percentage. What formula can you enter in cell **G3** to compute the percentage of the population living in urban areas?

4. The following stacked area chart causes students lots of problems in interpretation. In 2010 the urban section hits about 250 million while the top part hits around 310 million. What do these numbers represent?

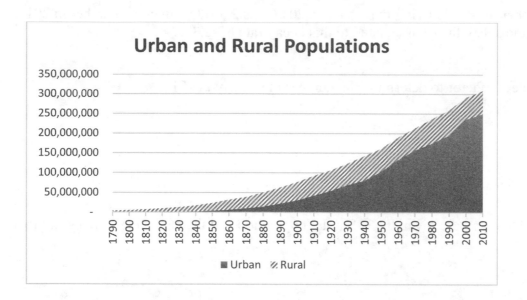

Section 4.3 Reflections
Monitor Your Understanding

Percent change is something many college students cannot correctly compute. How comfortable are you calculating percent change?

Do you find growth factors helpful, or do you prefer the proportion technique?

Section 4.4 Data Tables with Percentages

Objective 1 – Interpret Percentage Points
Objective 2 – Use Two-Way Tables

Objective 1 – Interpret Percentage Points

Key Terms

Percentage points	

Summary

Percent change requires extra caution when dealing with a change in percentages as opposed to numerical amounts. The total change between two percentages is given in units of percentage points (pp). If the poverty rate dropped from 12.7% to 12.3% from 2016 to 2017, then this is a 0.4 percentage point drop giving a -3.1% change. We can't say it dropped 0.4% because this would confuse it with the -3.1%.

As you read, jot down notes and questions. At the end of this section's Guided Worksheets there is a space for **Reflection** *and* **Monitoring Your Understanding***, where you can try and answer your questions after completing the guided activities and homework.*

Notes & Questions

Guided Practice Activity #1 – The Employment Situation November 2018

The Bureau of Labor Statistics publishes a monthly employment situation report. For November 2018[4] we have the following data:

Summary table A. Household data, seasonally adjusted
[Numbers in thousands]

Category	Nov. 2017	Sept. 2018	Oct. 2018	Nov. 2018	Change from: Oct. 2018- Nov. 2018
Employment status					
Civilian noninstitutional population..	255,949	258,290	258,514	258,708	194
Civilian labor force...	160,533	161,926	162,637	162,770	133
Participation rate...	62.7	62.7	62.9	62.9	0.0
Employed...	153,917	155,962	156,562	156,795	233
Employment-population ratio......................................	60.1	60.4	60.6	60.6	0.0
Unemployed...	6,616	5,964	6,075	5,975	-100
Unemployment rate..	4.1	3.7	3.7	3.7	0.0
Not in labor force...	95,416	96,364	95,877	95,937	60

---

[4] https://www.bls.gov/news.release/pdf/empsit.pdf

Ok first some questions to see if we understand the table.

1. How big was the civilian labor force in November 2018?

2. What percentage of the civilian non-institutional population was employed in November 2018?

3. What percentage of the civilian non-institutional population was participating in the labor force in November 2018?

4. The unemployment rate is a percentage of what group?

5. What was the total and percent change in the unemployment rate from November 2017 to 2018?

6. The following table gives unemployment rates by education level:

	Nov 2018	1 month change	1 year change
Less than high school	5.6%	- 0.4 pps	+ 0.4 pps
High school	3.5	- 0.5	- 0.8
Some college or associate degree	3.1	+ 0.1	- 0.5
Bachelor's degree or higher	2.2	+ 0.2	+ 0.1

a. Does the data show that 5.6% of the unemployed had less than a high school education or that 5.6% of those with less than a high school education were unemployed? Why?

b. What was the unemployment rate for those with a high school diploma in November 2017?

c. What was the percent change in unemployment rate for those with a high school diploma from November 2017 to November 2018?

d. What was the total and percent difference in unemployment rates for those with some college and those with bachelor's degree or higher in November 2018? Choose the order so you compute a negative change and use your percent change in a sentence.

Objective 2 – Use Two-Way Tables

Key Terms	
Two-way table	
Base	
Categorical variable	

Summary

When encountering a **percentage,** the first question one should ask is: *Percentage of what?* The "what" is known as the **base** of a percentage. For example, why are the percentages declining but not the number of literary readers in **Table 4.14**? Are the bases of each percentage the same?

TABLE 4.14 Decline in Literary Reading

	1982	1992	2002	2015
% of U.S. Adult Population Reading Literature	56.9	54.0	46.7	43.1
Number of Literary Readers (in millions)	96	100	96	104

Data from: Decline in Literary Reading from *Reading at Risk*, June 2004, NEA, http://arts.gov/publications/reading-risk-survey-literary-reading-america-0. Also Annual Arts Basic Survey (AABS) NEA 2015.

In 1982 we are given that 56.9% (96 million) of U.S. adults, x, were literary readers

$$56.9\% \cdot x = 96$$

$$x = \frac{6}{0.569} = 168.7 \text{ million}$$

Similar calculations for 1992, $x = \frac{100}{0.54} = 185.2$ million, 2002, $x = \frac{96}{0.467} = 205.6$ million, and

2015, $x = \frac{104}{0.431} = 241.3$ million, show we are dealing with different populations and therefore give the different **bases of the percentages**. We are actually taking a smaller percentage of a growing population, resulting in relatively flat number of literary readers. Understanding the **base of a percentage** is intimately related to **two-way tables**, which we use when our data involves two **categorical variables**, like gender or political affiliation.

As you read, jot down notes and questions. At the end of this section's Guided Worksheets there is a space for
Reflection *and* ***Monitoring Your Understanding****, where you can try and answer your questions after completing the guided activities and homework.*

Notes & Questions

Guided Practice Activity #1 – Poverty Rates

The Census Bureau Report[5] *Income and Poverty in the U.S. 2017* gives the following data on poverty by age group (thousands of persons) with % of total population:

Age Group	Poor	Non-Poor	Total
< 18	12,808	60,548	73,356
	4.0%	18.8%	22.7%
18-64	22,209	175,904	198,113
	6.9%	54.5%	61.4%
65+	4,681	46,399	51,080
	1.5%	14.4%	15.8%
Total	39,698	282,851	322,549
	12.3%	87.7%	100.0%

1. What percentage of the poor are children?

2. Use the 4.0% from the table in a sentence.

3. What percentage of the children are poor?

4. What percentage of elderly are non-poor?

5. What percentage of the population is elderly?

---

[5] https://www.census.gov/content/dam/Census/library/publications/2018/demo/p60-263.pdf

6. Use the 87.7% from the table in a sentence.

7. What is the poverty rate per 100,000?

Excel Preview – 4.1 Population

In this excel problem you are given the U.S. population by age and sex for 2015 and estimated for 2060.

	A	B	C	D	E	F
1						
2	Table 3. Percent Distribution of the Projected Population by Selected Age Groups and Sex for the United State					
3						
4						
5				a.)	b.)	b.)
6	Sex and age			(Percent of total resident population as of July 1)		
7		2015	2060	Total Change	Percent Change	Percent Change
8	BOTH SEXES	100.00	100.00	(pp's)	(decimal)	(percentage)
9	Under 18 years	23.19	21.25			
10	Under 5 years	6.55	5.89			
11	5 to 13 years	11.44	10.65			
12	14 to 17 years	5.20	4.71			
13	18 to 64 years	61.97	56.86			
14	18 to 24 years	9.64	8.38			
15	25 to 44 years	26.24	25.29			
16	45 to 64 years	26.09	23.18			
17	65 years and over	14.84	21.90			
18	85 years and over	1.96	4.33			
19	100 years and over	0.02	0.16			
20	16 years and over	79.40	81.11			
21	18 years and over	76.81	78.75			
22	15 to 44 years	39.78	37.21			

1. To compute the total change in the percent of resident population between 2015 and 2060, what formula would you type into cell **D9** that can be filled down?

2. Why are the units pp's?

3. What formula would you enter in cell **E9** to compute the percent change that can be filled down?

4. In Excel we don't multiply by 100% to change a decimal into a percentage (try it!). We simply format the decimal as a percentage. What formula would you enter into cell **F9** that can be filled down?

5. How do you format a cell as a percentage?

6. Notice all the age groups in column **A**. What three age groups make up the entire population with no overlap between the three groups?

Section 4.4 Reflections
Monitor Your Understanding

Can you explain when percentage points are used and why they are necessary?

| **Objective 1** – Understand Percentiles |
| **Objective 2** – Use Frequency Polygons |

| **Objective 1** – Understand Percentiles |

Key Terms

Percentile	
Boxplot	
Five number summary	
Inter-quartile range	
Outliers	

Summary

In the Chapter 3 *Spotlight on Statistics* we looked at **percentages** associated with the **normal distribution** and heights of U.S. women. In Figure 4.3 we can see two different scales on the *x*-axis. First we are given *z*-scores (or number of **standard deviations** from the **mean**), and second we are given the cumulative percent of data up to each **z-score**. Note that percentiles and *z*-scores are "measures of position", meaning they tell us where a data value is located relative to the other data. To be at the 90th percentile means the data value is greater than 90% of the data.

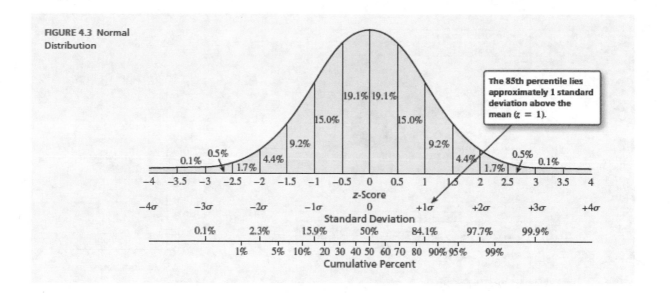

FIGURE 4.3 Normal Distribution

Boxplots give a basic visualization of your data's distribution relative to the quartiles.

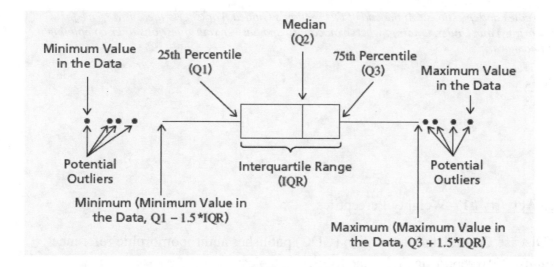

Boxplots and **histograms** are the two most common visualizations of a data set, both giving a different view of the distribution of your data.

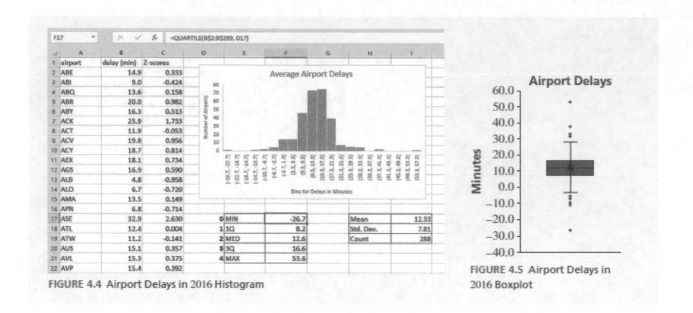

FIGURE 4.4 Airport Delays in 2016 Histogram

FIGURE 4.5 Airport Delays in 2016 Boxplot

As you read, jot down notes and questions. At the end of this section's Guided Worksheets there is a space for **Reflection** *and* **Monitoring Your Understanding***, where you can try and answer your questions after completing the guided activities and homework.*

Notes & Questions

Guided Practice Activity #1 – Weighty Percentiles

The Centers for Disease Control and Prevention (CDC) publishes anthropomorphic reference data and their August 2016 report[6] includes the following tables:

Table 4. Weight in pounds for females aged 20 and over and number of examined persons, mean, standard error of the mean, and selected percentiles, by race and Hispanic origin and age: United States, 2011–2014

Race and Hispanic origin and age	Number of examined persons	Mean	Standard error of the mean	Percentile								
				5th	10th	15th	25th	50th	75th	85th	90th	95th
All racial and Hispanic-origin groups[1]				Pounds								
20 years and over	5,425	168.5	0.92	110.5	119.8	126.8	136.0	159.1	191.1	212.2	229.3	256.8
20–29 years	853	161.8	1.88	107.2	113.9	119.7	128.8	149.8	184.6	212.8	228.7	255.2
30–39 years	915	172.9	1.92	114.9	124.0	128.7	137.7	160.5	195.9	218.1	237.7	269.2
40–49 years	979	173.1	2.21	115.6	124.8	130.7	138.8	164.0	196.8	217.7	237.2	261.3
50–59 years	923	174.4	2.31	115.2	124.9	129.2	141.0	167.4	197.2	219.4	239.4	271.3
60–69 years	889	168.8	1.98	107.9	122.5	129.6	140.3	164.4	191.2	206.8	224.8	245.6
70–79 years	527	165.8	2.08	108.2	120.2	128.1	138.6	158.5	188.5	205.8	216.5	238.2
80 years and over	339	141.9	1.64	96.9	103.8	110.7	121.7	141.0	160.6	172.4	180.4	190.6

---

[6] Fryar CD, Gu Q, Ogden CL, Flegal KM. Anthropometric reference data for children and adults: United States, 2011–2014. National Center for Health Statistics. Vital Health Stat 3(39). 2016.

Table 6. Weight in pounds for males aged 20 and over and number of examined persons, mean, standard error of the mean, and selected percentiles, by race and Hispanic origin and age: United States, 2011–2014

Race and Hispanic origin and age	Number of examined persons	Mean	Standard error of the mean	Percentile								
				5th	10th	15th	25th	50th	75th	85th	90th	95th
All racial and Hispanic-origin groups[1]							Pounds					
20 years and over	5,236	195.7	0.94	136.7	146.2	154.2	165.1	189.3	218.8	236.7	249.9	275.4
20–29 years	936	186.8	2.60	126.2	137.6	143.9	152.9	177.8	208.5	232.1	247.2	280.8
30–39 years	914	198.8	1.73	140.2	150.2	158.2	168.1	190.8	221.2	242.8	259.6	281.9
40–49 years	872	201.7	1.60	146.2	156.3	162.9	171.7	196.4	222.5	237.2	249.0	279.2
50–59 years	854	199.5	2.03	140.0	152.0	160.0	170.2	195.9	222.3	236.9	250.4	279.4
60–69 years	874	199.7	3.02	137.9	147.0	154.3	168.0	195.3	223.1	244.2	255.3	279.2
70–79 years	486	189.3	2.03	136.5	146.2	152.6	166.6	183.6	212.0	227.0	236.3	251.5
80 years and over	300	174.6	1.90	125.2	132.6	141.4	154.2	171.1	194.5	207.1	216.1	233.5

1. What percentage of data for men 20 years and over lies between 146.2 lbs and 218.8 lbs?

2. What is the 25th percentile for women 20 years and over?

3. What are the middle three quartiles for women 20 years and over?

4. Identify all given deciles for men 20 years and over.

5. Identify all given quintiles for women.

6. What is the probability of a random male 20 years and over having a weight greater than 236.7 lbs?

Objective 2 – Use Frequency Polygons

Key Terms

Relative frequency	
Cumulative frequency	
Relative frequency polygon	
Cumulative frequency polygon	

Summary

Recall that the normal distribution in Figure 4.3 above is just a histogram with the percentages giving the number of data values in bins. Figure 4.4 shows a histogram giving the relative frequencies of words of varying lengths from samples of Mark Twain's writings, with data given in Table 4.22.

Note that the relative frequencies are just percentages of the total number of words, and the cumulative frequencies simply sum the relative frequencies up to each word length (i.e. they are percentiles).

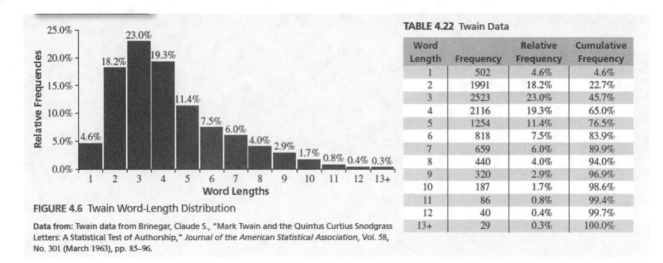

FIGURE 4.6 Twain Word-Length Distribution

Data from: Twain data from Brinegar, Claude S., "Mark Twain and the Quintus Curtius Snodgrass Letters: A Statistical Test of Authorship," *Journal of the American Statistical Association*, Vol. 58, No. 301 (March 1963), pp. 85–96.

TABLE 4.22 Twain Data

Word Length	Frequency	Relative Frequency	Cumulative Frequency
1	502	4.6%	4.6%
2	1991	18.2%	22.7%
3	2523	23.0%	45.7%
4	2116	19.3%	65.0%
5	1254	11.4%	76.5%
6	818	7.5%	83.9%
7	659	6.0%	89.9%
8	440	4.0%	94.0%
9	320	2.9%	96.9%
10	187	1.7%	98.6%
11	86	0.8%	99.4%
12	40	0.4%	99.7%
13+	29	0.3%	100.0%

If instead of making a histogram of the relative frequencies we make a line graph, we get what is called a relative frequency polygon. Graphing the cumulative frequencies yields a cumulative frequency polygon. Note that percentiles are essentially cumulative frequency cutoffs.

National Vital Statistics Reports, Vol. 67, No. 1, January 31, 2018

As you read, jot down notes and questions. At the end of this section's Guided Worksheets there is a space for **Reflection** *and* **Monitoring Your Understanding***, where you can try and answer your questions after completing the guided activities and homework.*

Notes & Questions

Guided Practice Activity #1 – Birth Weighty Percentiles

The Centers for Disease Control and Prevention produce national vital statistics reports that include the following data on birthweights:

Table 22. Births, by birthweight (grams): United States, 2010–2016, and by race and Hispanic origin of mother, 2016

Year and race and Hispanic origin of mother	All births	Total under 2,500	Total under 1,500	Less than 500	500–999	1,000–1,499	1,500–1,999	2,000–2,499	2,500–2,999	3,000–3,499	3,500–3,999	4,000–4,499	4,500–4,999	5,000 or more	Not stated
All races and origins¹	Number						Percent								Number
2016	3,945,875	8.17	1.40	0.14	0.52	0.74	1.59	5.17	18.54	38.76	26.60	6.85	0.97	0.11	4,518
2015	3,978,497	8.07	1.40	0.15	0.52	0.73	1.58	5.09	18.36	38.84	26.73	6.90	0.98	0.12	3,621
2014	3,988,076	8.00	1.40	0.15	0.52	0.74	1.56	5.04	18.27	38.80	26.88	6.94	0.99	0.12	3,270
2013	3,932,181	8.02	1.41	0.15	0.53	0.73	1.56	5.05	18.22	38.93	26.85	6.86	0.99	0.11	4,452
2012	3,952,841	7.99	1.42	0.15	0.54	0.73	1.56	5.01	18.28	39.00	26.81	6.83	0.97	0.12	4,008
2011	3,953,590	8.10	1.44	0.15	0.54	0.75	1.58	5.08	18.44	39.13	26.56	6.71	0.95	0.11	4,570
2010	3,999,386	8.15	1.45	0.15	0.55	0.75	1.59	5.11	18.63	39.21	26.41	6.58	0.92	0.11	3,964
2016															
Non-Hispanic, single race²:															
White	2,056,332	6.97	1.07	0.09	0.37	0.61	1.39	4.51	15.88	37.59	29.82	8.39	1.21	0.12	2,212
Black	558,622	13.68	2.95	0.37	1.20	1.38	2.67	8.06	25.38	38.24	18.42	3.69	0.52	0.07	936
Hispanic³	918,447	7.32	1.24	0.13	0.46	0.65	1.41	4.68	18.76	40.79	26.06	6.10	0.85	0.12	556
All races and origins¹							Number								
2016	3,945,875	321,839	55,110	5,710	20,323	29,077	62,863	203,866	730,710	1,527,707	1,048,476	269,865	38,264	4,496	4,518
2015	3,978,497	320,869	55,592	5,863	20,689	29,040	62,862	202,415	729,673	1,544,024	1,062,456	274,404	38,796	4,654	3,621
2014	3,988,076	318,847	55,947	5,936	20,721	29,290	61,992	200,908	727,987	1,546,274	1,071,007	276,592	39,353	4,746	3,270
2013	3,932,181	315,099	55,458	5,945	20,866	28,647	61,238	198,403	715,764	1,529,258	1,054,767	269,594	38,834	4,416	4,449
2012	3,952,841	315,709	56,252	5,947	21,432	28,873	61,499	197,958	721,840	1,540,161	1,058,604	269,581	38,288	4,650	4,008
2011	3,953,590	319,711	56,754	5,942	21,289	29,523	62,504	200,453	728,201	1,545,355	1,048,902	265,040	37,475	4,336	4,570
2010	3,999,386	325,563	57,841	5,980	22,015	29,846	63,427	204,295	744,181	1,566,755	1,055,004	262,997	36,706	4,216	3,964
2016															
Non-Hispanic, single race²:															
White	2,056,332	143,254	21,979	1,888	7,565	12,526	28,578	92,697	326,279	772,165	612,641	172,434	24,781	2,566	2,212
Black	558,622	76,299	16,465	2,083	6,675	7,707	14,885	44,949	141,557	213,260	102,702	20,571	2,880	417	936
Hispanic³	918,447	67,210	11,378	1,175	4,237	5,966	12,910	42,922	172,171	374,434	239,210	55,983	7,770	1,113	556

¹Includes births to race and origin groups not shown separately, such as Hispanic single-race white, Hispanic single-race black, and non-Hispanic multiple-race women, as well as births with origin not stated.
²Race and Hispanic origin are reported separately on birth certificates; persons of Hispanic origin may be of any race. In this table, non-Hispanic women are classified by race. Race categories are consistent with 1997 Office of Management and Budget standards; see Technical Notes. Single race is defined as only one race reported on the birth certificate.
³Includes all persons of Hispanic origin of any race; see Technical Notes.

NOTE: Equivalents of gram weights in pounds and ounces are shown in Technical Notes.
SOURCE: NCHS, National Vital Statistics System, Natality.

1. Use the number 6.97 from Total Under 2500/White in a meaningful sentence.

2. What is the frequency from the table associated with the relative frequency, 6.97, from #1?

3. 6.97% of what number gives the frequency in #2? Note that computing the frequency on your own gives a slightly different answer from the table due to round-off error.

Entering some of this data into Excel we can make a histogram (of sorts).

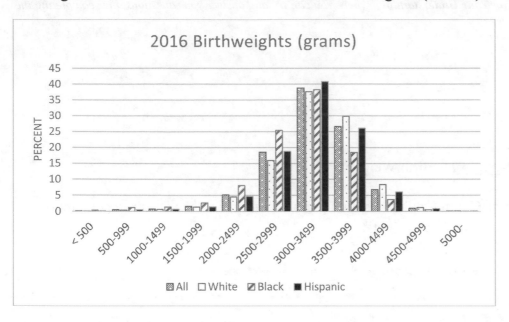

4. What demographic group seems to have low birthweight babies?

5. Use the relative frequencies to compute the cumulative frequencies and fill in the table:

2016 Birthweights (grams)				
Relative Frequencies				
	All	White	Black	Hispanic
< 500	0.14	0.09	0.37	0.13
500-999	0.52	0.37	1.2	0.46
1000-1499	0.74	0.61	1.38	0.65
1500-1999	1.59	1.39	2.67	1.41
2000-2499	5.17	4.51	8.06	4.68
2500-2999	18.54	15.88	25.38	18.76
3000-3499	38.76	37.59	38.24	40.79
3500-3999	26.6	29.82	18.42	26.06
4000-4499	6.85	8.39	3.69	6.1
4500-4999	0.97	1.21	0.52	0.85
5000-	0.11	0.12	0.07	0.12

2016 Birthweights (grams)				
Cumulative Frequencies				
	All	White	Black	Hispanic
< 500				
500-999				
1000-1499				
1500-1999				
2000-2499				
2500-2999				
3000-3499				
3500-3999				
4000-4499				
4500-4999				
5000-				

<u>**Excel Preview**</u> – Birthweights

This Excel problem is not in the homework but uses the birthweight information above to make cumulative frequency polygons.

	A	B	C	D	E	F	G	H	I	J	K
1	2016 Birthweights (grams)						2016 Birthweights (grams)				
2		Relative Frequencies						Cumulative Frequencies			
3		All	White	Black	Hispanic			All	White	Black	Hispanic
4	< 500	0.14	0.09	0.37	0.13		< 500	0.14	0.09	0.37	0.13
5	500-999	0.52	0.37	1.2	0.46		500-999				
6	1000-1499	0.74	0.61	1.38	0.65		1000-1499				
7	1500-1999	1.59	1.39	2.67	1.41		1500-1999				
8	2000-2499	5.17	4.51	8.06	4.68		2000-2499				
9	2500-2999	18.54	15.88	25.38	18.76		2500-2999				
10	3000-3499	38.76	37.59	38.24	40.79		3000-3499				
11	3500-3999	26.6	29.82	18.42	26.06		3500-3999				
12	4000-4499	6.85	8.39	3.69	6.1		4000-4499				
13	4500-4999	0.97	1.21	0.52	0.85		4500-4999				
14	5000-	0.11	0.12	0.07	0.12		5000-				
15											

1. In the guided activity above you were asked to compute all of the cumulative frequencies by hand. Hopefully you will now appreciate using Excel to do this task! What formula can you enter in cell **H5** than can be filled down **and** across to compute all of the cumulative frequencies for you? Note this is tricky but can be done, think about it! Seriously you can do this :O)

2. We can then create a cumulative frequency polygon. Which demographic group has their polygon to the left of the others and what does this indicate?

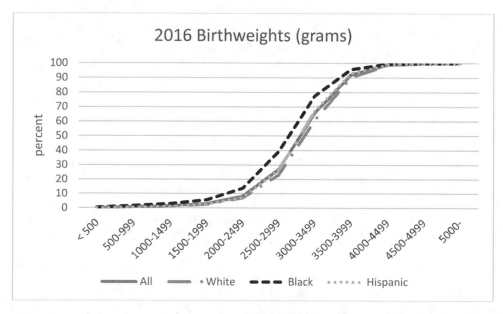

Section 4.5 Reflections
Monitor Your Understanding

Hopefully you were able to figure out the formula in the Excel activity above and had a "Eureka!" moment. I love those. Describe your thought process and how you solved the problem.

Chapter 5 – Linear Functions

Section 5.1 Proportionality

> **Objective 1** – Explore Proportionality in Tables and Graphs
> **Objective 2** – Explore Proportionality in Equations

> **Objective 1** – Explore Proportionality in Tables and Graphs

Summary

We studied **functions** in Chapter 1 and saw how prevalent and useful the study of relationships between quantities can be. In Chapter 2, we introduced **ratios** with **proportionality** as the most basic of functional relationships. In this chapter we will study another aspect of proportionality, the fact that the graphical representation of a proportional relationship is linear.

Table 5.2 shows the cost of monthly TracFone plans as proportional to the number of minutes talked at **rates** of 5¢ and 7¢ per minute. From Table 5.2 we created the graph (*x-y* scatterplot) in Figure 5.2 which gives us a visual display of the relationships between the inputs and outputs of the functions, noting that the points for each function all fall on a line starting at the origin. Increasing the input by 1 minute necessitates an increase of 5 cents in the first output. The rate, 5¢ per minute, thus gives a measure of how steep the line is; the plan that charges 7¢ per minute has a steeper line (i.e. greater rate of incline). The rate is therefore called the slope of the line and will be formally defined shortly.

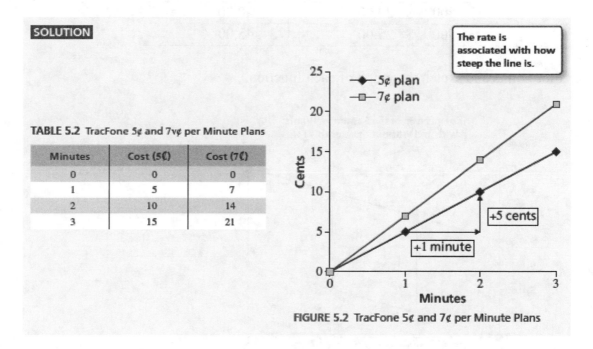

SOLUTION

The rate is associated with how steep the line is.

TABLE 5.2 TracFone 5¢ and 7v¢ per Minute Plans

Minutes	Cost (5¢)	Cost (7¢)
0	0	0
1	5	7
2	10	14
3	15	21

+5 cents

+1 minute

FIGURE 5.2 TracFone 5¢ and 7¢ per Minute Plans

As you read, jot down notes and questions. At the end of this section's Guided Worksheets there is a space for **Reflection** *and* **Monitoring Your Understanding**, *where you can try and answer your questions after completing the guided activities and homework.*

Notes & Questions

Guided Practice Activity #1 – Proportional Tables and Scatterplots

This worksheet connects the concepts of proportionality and constant rate to linear functions. Please read through the first example that has been done for you, and then complete the following 3 examples.

1. **Monthly cell phone bill**: Cost is proportional to minutes you speak at a rate of 5 cents per minute. Next assume there is an additional fixed cost of $20 each month.

   a. Fill in table of values for both functions:

Minutes	Cost	Cost + $20
0	$ -	$ 20.00
100	$ 5.00	$ 25.00
200	$ 10.00	$ 30.00
300	$ 15.00	$ 35.00
400	$ 20.00	$ 40.00
500	$ 25.00	$ 45.00

   b. Plot points and sketch the graph of each function.

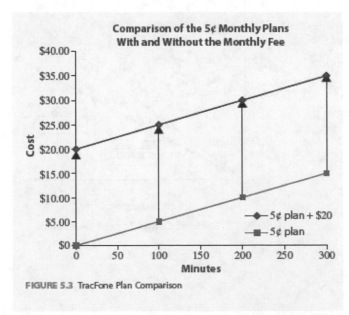

FIGURE 5.3 TracFone Plan Comparison

2. **Weight of water:** The weight of water is proportional to the volume at a rate of 8 pounds per gallon. Next assume the bucket you are using to carry the water weighs 10 pounds.

    a. Fill in table of values for both functions:

V (gal)	W (lbs)	W(lbs) + 10 lbs.
0		
0.5		
1.0		
1.5		
2.0		
2.5		

    b. Plot points/ sketch graph of both functions. Be sure to label axes:

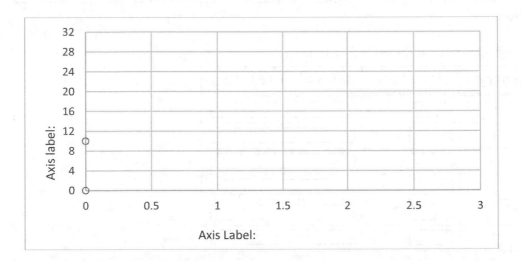

3. **Buying tickets:** The cost is proportional to the number of tickets you buy at the fair at a rate of $2 per ticket. Next assume there is a fixed cost of $15 to get into the fair.

    a. Fill in table of values for both functions:

T	C ($)	C ($) + $15
0		
5		
10		
15		
20		
25		

b. Plot points and sketch the graph of each function:

Cost of Tickets

4. **Driving distance:** The distance you drive is proportional to the time driving at a constant speed of 40 miles per hour. Next assume the distance driven includes the 100 miles you already drove.

   a. Fill in table of values for both functions:

T (hours)	D (miles)	D (miles) + 100 miles
0		
1		
2		
3		
4		
5		

   b. Plot points and sketch graph of each function.

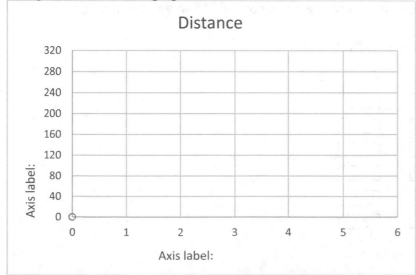

Distance

Objective 2 – Explore Proportionality in Equations

Summary

We have just seen that the **constant of proportionality** is a **rate** which determines the steepness of the line that graphs the **proportional** relationship. Note that the lines go through the origin for proportional quantities, and you should be able to convince yourself that any line through the origin implies the input and output are proportional. Recall from Chapter 3 that all proportional relationships can be represented with an equation of the form $y = k \cdot x$:

$$\text{Cost} = 5 \cdot \text{Minutes} \qquad \text{Cost} = 7 \cdot \text{Minutes}$$
$$C = 5 \cdot M \qquad\qquad C = 7 \cdot M$$

Each rate is the associated **constant of proportionality** in the equation.

We also saw that adding in a monthly fee to the Tracfone example simply raised the linear graph. To find the equation of the monthly plan with a $20 monthly fee, our cost starts at $20 and we add in the 5¢ per minute:

$$\text{Cost} = \$20.00 + \$0.05 \cdot \text{Minutes}$$
$$C = \$0.05 \cdot M + \$20$$

Note that this is not the equation of a proportional relationship, $y = k \cdot x$.

CAUTION! Equations are typically written with x and y,

$$y = 0.05x + 20$$

but you must use care in determining what each variable represents and what the units are.

As you read, jot down notes and questions. At the end of this section's Guided Worksheets there is a space for **Reflection** *and* **Monitoring Your Understanding**, *where you can try and answer your questions after completing the guided activities and homework.*

Notes & Questions

Guided Practice Activity #1 – Proportional Equations

This worksheet connects the concepts of proportionality and constant rate to linear functions. Please read through the first example that has been done for you, and then complete the following 3 examples.

1. **Monthly cell phone bill**: Cost is proportional to minutes you speak at a rate of 5 cents per minute.

    a. Write an equation for this proportional relationship.
        $$\text{Cost} = 0.05 \cdot \text{Minutes}$$

    b. Now assume there is an additional fixed cost of $20 each month. Write down the new equation.
        $$\text{Cost}_2 = 0.05 \cdot \text{Minutes} + 20$$

2. **Weight of water:** The weight of water is proportional to the volume at a rate of 8 pounds per gallon.

    a. Write an equation for this proportional relationship.

    b. Now assume the bucket you are using to carry the water weighs 10 pounds. Write down the new equation.

3. **Buying tickets:** The cost is proportional to the number of tickets you buy at the fair at a rate of $2 per ticket.

    a. Write an equation for this proportional relationship.

    b. Now assume there is a fixed cost of $15 to get into the fair. Write down the new equation.

4. **Driving distance:** The distance you drive is proportional to the time driving at a constant speed of 40 miles per hour.

    a. Write an equation for this proportional relationship.

    b. Now assume the distance driven includes the 100 miles you already drove. Write down the new equation.

Excel Preview – 4.1 Lines

Spreadsheets are perfect for computing the *y*-values given an equation.

	A	B	C	D	E	F
1	This problem has you compare various linear functions.					
2						
3						
4		a.)	a.)	a.)	a.)	a.)
5	Inputs (*x*)			Outputs (*y*)		
6	*x*	$y = -5x + 3$	$y = -2x + 3$	$y = 0x + 3$	$y = 2x + 3$	$y = 5x + 3$
7	-10	53	23	3	-17	-47
8	-9	48	21	3	-15	-42
9	-8	43	19	3	-13	-37
10	-7	38	17	3	-11	-32
11	-6	33	15	3	-9	-27

1. What formula is in cell **B7** that can be filled down?

2. What change could you make to the spreadsheet to allow for a formula in cell **B7** that can be filled down and across?

3. In the following scatterplot label the different lines with their equations from the screenshot above.

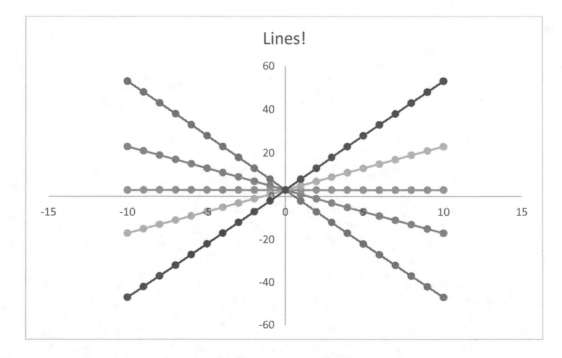

Lines!

<u>**Section 5.1 Reflections**</u>
Monitor Your Understanding

What have you learned about lines that was new or different?

Section 5.2 Slope and *y*-Intercepts

| **Objective 1** – Interpret Slopes as Rates |
| **Objective 2** – Interpret *y*-intercepts as Initial Values |

| **Objective 1** – Interpret Slopes as Rates |

Key Terms

Slope	

Summary

The TracFone examples have shown that the rate associated with a proportional relationship (i.e. 5¢ per minute) has a geometric interpretation as the steepness or slope of the line we get when we graph the function. The linearity of the graph is captured by the constant of proportionality. In general, given any two points we can draw a line connecting them. The steepness of this line is measured by the ratio of the total change in *y*-values to the total change in *x*-values.

CAUTION! When interpreting the slope as a rate, think:

The <output> is changing by <slope> <y-units> per <x-units>.

For example:

The **cost** *is changing by* **0.05 dollars** *per* **minute**.

The notation Δx and Δy indicates change in *x* and change in *y*. The slope is thus the ratio of change in *y* to change in *x*.

As you read, jot down notes and questions. At the end of this section's Guided Worksheets there is a space for **Reflection** *and* **Monitoring Your Understanding***, where you can try and answer your questions after completing the guided activities and homework.*

Notes & Questions

Guided Practice Activity #1 – Slopes

This worksheet asks you to interpret slopes as rates using the sentence template:

The <output> is changing by <slope> <y-units> per <x-units>.

1. Monthly cell phone bill:

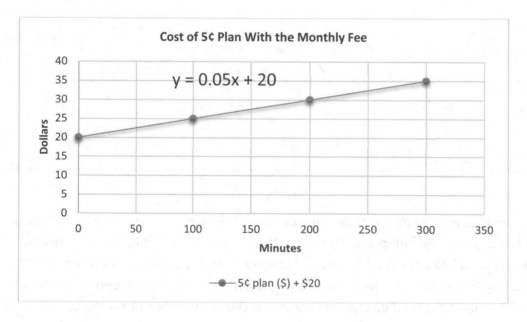

a. Interpret the slope of the equation in real world terms.

2. **Concentration of pollutants:**

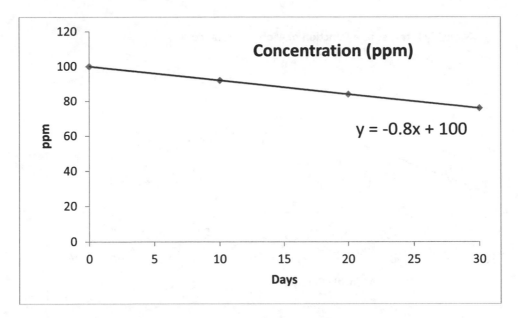

a. Interpret the slope of the equation in real world terms.

3. **Houston population:**

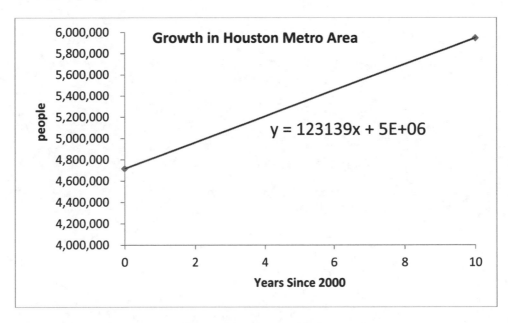

a. Interpret the slope of the equation in real world terms.

4. Monthly Interest:

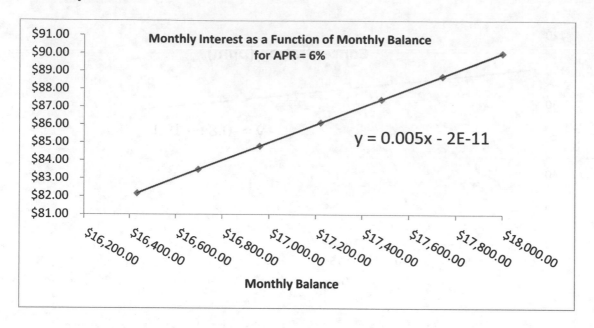

Monthly Interest as a Function of Monthly Balance for APR = 6%

$y = 0.005x - 2E-11$

a. Interpret the slope of the equation in real world terms.

Key Terms

y-Intercept	

Summary

Every line has two key characteristics, it's steepness, as measured by the **slope**, and its location. Recall the TracFone examples where we added $20 as a monthly fee which raised the graph of the line up 20 units, but did not change the slope. We characterize the location of the line by specifying where it intercepts the *y*-axis, i.e. the *x*-coordinate is zero.

If your input is time then the *x*-coordinate is zero can be thought of as the initial value or starting point. Unfortunately if you have time-series data where the inputs are years (1990, 1995, etc.) then $x = 0$ refers to 0 C.E., and the *y*-intercept will have no relevance to today (the U.S. did not exist back then for example!). This is the one time in this course when the correct answer may very well be: "This makes no sense!" All other times it should make sense and you need to keep working :O). So great care must be taken when interpreting the *y*-intercept, and we often measure time from a more recent year, translating our units to something like *years since 1990*.

*As you read, jot down notes and questions. At the end of this section's Guided Worksheets there is a space for **Reflection** and **Monitoring Your Understanding**, where you can try and answer your questions after completing the guided activities and homework.*

Notes & Questions

<u>**Guided Practice Activity #1**</u> – *y*-intercepts

This worksheet asks you to interpret *y*-intercepts.

1. **Monthly cell phone bill**:

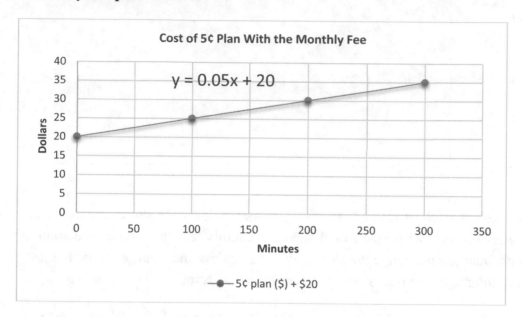

a. Interpret *y*-intercept of the equation in real world terms.

2. **Concentration of pollutants:**

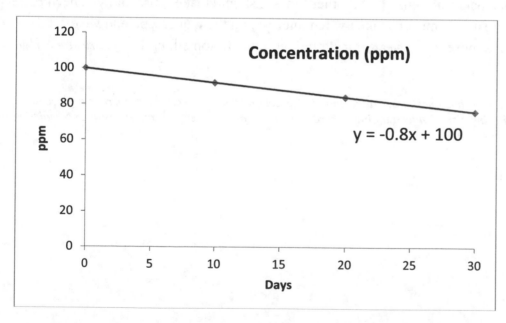

a. Interpret the *y*-intercept of the equation in real world terms.

3. Houston population:

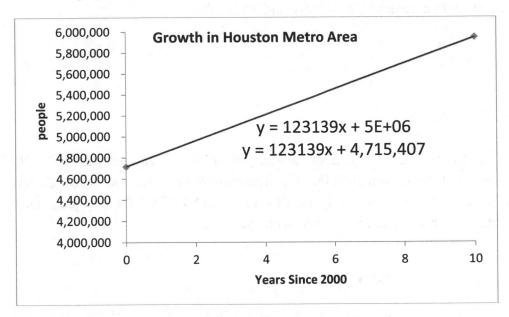

a. Interpret the *y*-intercept of the equation in real world terms. Note that we have two versions of the equation, please interpret both numbers.

4. Monthly Interest:

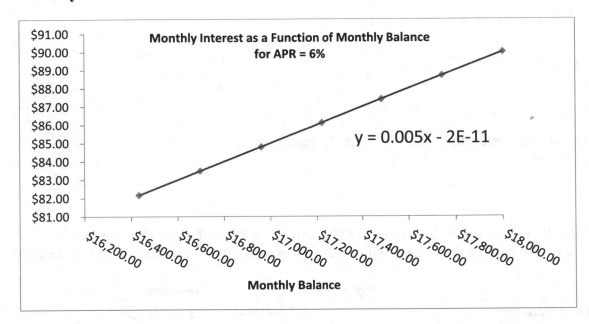

a. Interpret the *y*-intercept of the equation in real world terms. Be sure to convert the scientific notation to a number that is more understandable.

Excel Preview – 4.2 Population

This problem asks you to first graph the U.S. populations given from every census, 1790 – 2010, and then to compare to just those from 1950 on. It is important to pay attention to how Excel rounds off values in the equations, and how to use **SLOPE** and **INTERCEPT** for the actual values. Also important to understand how to deal with "years since".

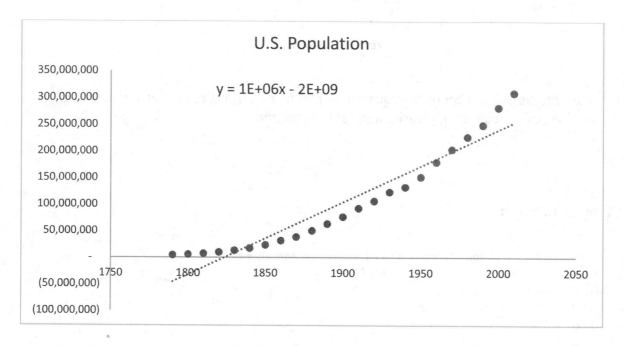

1. Interpret the slope and *y*-intercept from the equation.

2. Using the Excel built-in functions, **SLOPE** and **INTERCEPT**, we can compute the actual values. Use both pairs of slope and intercept to estimate the population of the U.S. in 2050.

	Slope:	Intercept:
From Equation:	1,000,000	-2,000,000,000
Using SLOPE and INTERCEPT:	1,360,370	-2,480,724,026

3. Finally we can make a scatterplot of the population using years since 1950, and of course use **SLOPE** and **INTERCEPT** for the accurate values. Wow that looks linear!

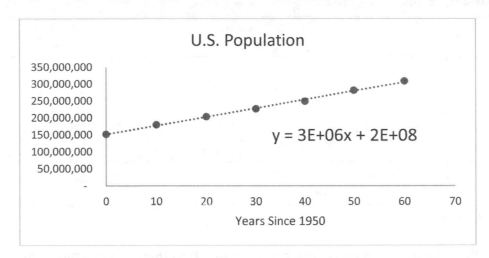

U.S. Population

$y = 3E+06x + 2E+08$

Years Since 1950

	Slope:	Intercept:
From Equation:	3,000,000	200,000,000
Using SLOPE and INTERCEPT:	2,578,409	151,116,864

a. Interpret the **SLOPE** and **INTERCEPT** values.

b. Use the accurate values to predict the U.S population 2050.

c. Which estimate for 2050 is more accurate, from **#2** or **#3**? Why?

Section 5.2 Reflections
Monitor Your Understanding

Objective 1 – Evaluate the Slope-Intercept Form

Key Terms

Linear function	
Slope-intercept form	

Summary

We have encountered three equations in this chapter representing **functions** with a graph that is a straight line:

1. $C = \$0.05 \cdot M + \$20.00$
2. $C = 100 - 0.8 \cdot d$
3. $I = 0.5\% \cdot B$

All three of these equations can be rewritten in the form, $y = m \cdot x + b$:

1. $y = 0.05 \cdot x + 20$
2. $y = -0.8 \cdot x + 100$
3. $y = 0.005 \cdot x + 0$

The coefficient of x is the **slope** of the line or the constant rate of total change between the input and output variables. The constant at the end of the equation is the **y-intercept** or y-value which occurs when $x = 0$.

As you read, jot down notes and questions. At the end of this section's Guided Worksheets there is a space for **Reflection** *and* **Monitoring Your Understanding***, where you can try and answer your questions after completing the guided activities and homework.*

Notes & Questions

Guided Practice Activity #1 – Metro Pops

The following table shows Palm Coast, FL as the fastest growing metropolitan area from 2000 – 2010.

TABLE 5.5 Population Change for the Ten Most Populous and Ten Fastest-Growing Metropolitan Statistical Areas: 2000 to 2010

Metropolitan statistical area	Population		Change	
	2000	2010	Number	Percent
MOST POPULOUS				
New York–Northern New Jersey–Long Island. NY-NJ-RA	18,323,002	18,897,109	574,107	3.1
Los Angeles-Long Beach-Santa Ana. CA	12,365,627	12,828,837	463,210	3.7
Chicago-Joliet-Naperville. IL-IN-WI	9,098,316	9,461,105	362,789	4.0
Dallas-Fort Worth–Arlington, TX	5,161,544	6,371,773	1,210,229	23.4
Philadelphia-Camden-Wilmington. PA-NJ-OE-MD	5,687,147	5,965,343	278,196	4.9
Houston-Sugar Land–Baytown. TX	4,715,407	5,946,800	1,231,393	26.1
Washington–Arlington-Alexandria. DC-VA-MD-WV	4,796,183	5,582,170	785,987	16.4
Miami–Fort Lauderdale-Pompano Beach, FL	5,007,564	5,564,635	557,071	11.1
Atlanta-Sandy Springs–Marietta, GA	4,247,981	5,268,860	1,020,879	24.0
Boston-Cambridge-Quincy. MA-NH	4,391,344	4,552,402	161,058	3.7
FASTEST-GROWING				
Palm Coast. FL	49,832	95,696	45,864	92.0
St. George. UT	90,354	138,115	47,761	52.9
Las Vegas–Paradise. NV	1,375,765	1,951,269	575,504	41.8
Raleigh-Cary. NC	797,071	1,130,490	333,419	41.8
Cape Coral–Fort Myers. FL	440,888	618,754	177,866	40.3
Provo-Orem. UT	376,774	526,810	150,036	39.8
Greeley, CO	180,926	252,825	71,899	39.7
Austin–Round Rock–San Marcos. TX	1,249,763	1,716,289	466,526	37.3
Myrtle Beach–North Myrtle Beach–Conway. SC	196,629	269,291	72,662	37.0
Bend. OR	115,367	57,733	42,366	36.7

Note: The full names of the metropolitan statistical areas are shown in this table: abbreviated versions of the names are shown in the text.
Data from: Largest and Fastest-Growing Cities, U.S. Census Bureau, 2010 Census, http://www.census.gov/prod/cen2010/briefs/c2010br-01.pdf

1. Use the values in the table to determine the linear equation of the form, $y = m \cdot x + b$, between the two points for Palm Coast, FL where x = years and y = population.

2. Interpret the slope and y-intercept.

3. Use the values in the table to determine the line between the two points for Palm Coast, FL where x = years and y = population.

4. Interpret the slope and *y*-intercept.

Excel Preview – 4.3 MF Incomes

This problem compares median incomes for males and females in the U.S. since 1947.

	A	B	C	D	E	F	G	H	I
1	Year			Male			Female		
2			Number with income (thous.)	Median income		Number with income (thous.)	Median income		
3				Current dollars	2016 $		Current dollars	2016 $	
4	UNITED STATES	a.)							b.)
5	Year	Years Since 1947			Male			Female	$100 in 2016
6	1947	0	46,813	2,230	20,968	21,479	1,017	9,563	$ 10.63
7	1948	1	47,370	2,396	20,860	22,725	1,009	8,785	$ 11.49
8	1949	2	48,258	2,346	20,680	23,510	960	8,462	$ 11.34
9	1950	3	47,585	2,570	22,375	24,651	953	8,297	$ 11.49
10	1951	4	47,497	2,952	23,819	25,179	1,045	8,432	$ 12.39

1. We are going to use years since 1947. What formula is in cell **B6** that can be filled down?

2. The data gives us median income in "current dollars" and "2016 $". Explain the difference between the two values, $1,017 and $9,563, for women in 1947.

3. We can use the values from **#2** to convert $100 in 2016 to 1947 $. What formula can be entered in cell **I6** that can be filled down?

4. The following scatterplot gives equations for men and women. Interpret both slopes and intercepts.

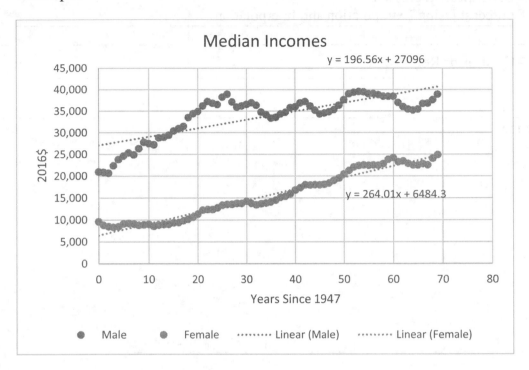

Median Incomes

y = 196.56x + 27096

y = 264.01x + 6484.3

2016$

Years Since 1947

● Male ● Female ·········· Linear (Male) ·········· Linear (Female)

5. Use both equations to predict when men and women will have equal median incomes.

In 305.6 years since 1947, so roughly mid 2252!

Section 5.3 Reflections
Monitor Your Understanding

Objective 1 – Use Linear Regression
Objective 2 – Forecast Using Extrapolation and Interpolation

Objective 1 – Use Linear Regression

Key Terms

Linear trendline	

Summary

In the real world, quantities do not usually vary in a perfectly linear fashion as in the pollutant **concentration** example or the population of Houston for more than two data points. The concentration levels or population would be measured regularly and then a *linear trend* might be observed, in which case a *linear model* would be constructed to help quantify and describe the way in which the concentration levels are dropping or the population is increasing. Consider, for example, the data on Social Security outlays (expenditures or costs) in the U.S.:

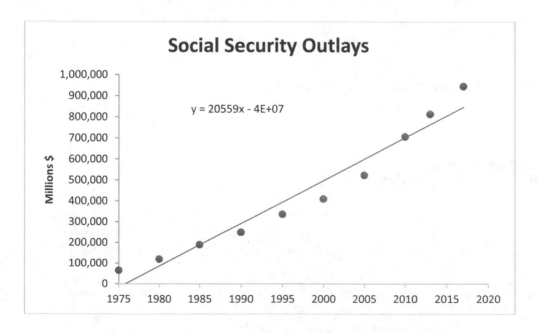

Note that Excel somehow chooses the "best fitting" line through the scatterplot, the algorithm behind this procedure will be discussed in Chapter 8. Excel has also given us the equation of the line,

$$y = 20{,}559 \cdot x - 4 \cdot 10^7$$

where we have interpreted the scientific notation, -4E+07, as a power of ten.

The **y-intercept**, -40,000,000, makes no sense as explained in the chapter. We include it here to show how Excel rounds off large numbers when displaying equations and to remind you to rescale your units of years as "*years since*" a certain year.

As you read, jot down notes and questions. At the end of this section's Guided Worksheets there is a space for **Reflection** *and* **Monitoring Your Understanding**, *where you can try and answer your questions after completing the guided activities and homework.*

Notes & Questions

Guided Practice Activity #1 – Social Security

The following table gives the amount of money spent on Social Security in this country:

Year	Years Since 1975	Social Security Outlays (millions $)
1975		64,658
1980		118,547
1985		188,623
1990		248,623
1995		335,846
2000		409,423
2005		523,305
2010		706,737
2013		813,551
2017		944,878
2020		

1. Take the first and last data points: (1975, $64,658) and (2017, $944,878) and compute the slope (1 decimal place). Interpret as a *rate*: something per something.

2. Draw the line between the two points on the following graph:

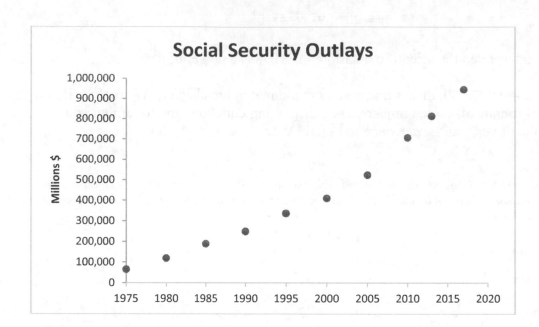

Social Security Outlays

Millions $

3. Compute the equation of the line: $y = mx + b$, by plugging in an (x, y) point and the slope, m, and solving for b.

4. Where does your line look like it will hit the vertical axis on the graph? What year does this vertical axis correspond to? How does this relate to the value of b in your equation in **#3**?

5. Now fill in the *Years Since 1975* column in the table and put the corresponding numbers on the *x*-axis of the graph above.

Year	Years Since 1975	Social Security Outlays
1975		64,658
1980		118,547
1985		188,623
1990		248,623
1995		335,846
2000		409,423
2005		523,305
2010		706,737
2013		813,551
2017		944,878
2020		

6. Compute the equation of the line through the points (0, $64,658) and (42, $944,878). How does this differ from the equation in **#3**?

7. Now consider the *trendline* (*best fit* line or *least-squares* line) in the graph below. How does this line differ from the one you drew?

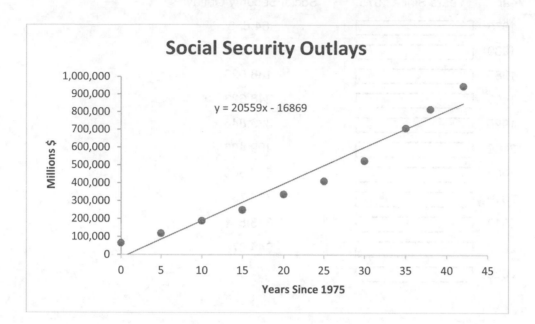

8. Interpret the slope and y-intercept in real world terms.

9. Use both your line (*x* = years since 1975) and the best fit line to predict Social Security Outlays in 2020.

10. Do you think the actual spending will be more or less than the estimates from **#9**?

Objective 2 – Forecast Using Extrapolation and Interpolation

Key Terms

Extrapolation	
Interpolation	

Summary

In the previous guided activity, we had Excel fit a **linear trendline** to Social Security data. Notice the line does not go through all or even most of the points in the scatterplot, and the **y-intercept** of -$16,689 million estimates the Social Security outlays in 1975 which we know to be $64,658 million. This may seem strange to you. Why estimate a known value?
The point is that this trendline captures the *trend* of Social Security Outlays and can now be used to make predictions in the future.

As you read, jot down notes and questions. At the end of this section's Guided Worksheets there is a space for **Reflection** *and* **Monitoring Your Understanding**, *where you can try and answer your questions after completing the guided activities and homework.*

Notes & Questions

Guided Practice Activity #1 – Divorce Rates

The following spreadsheet compares various divorce and marriage rates.

	A	B	C	D	E	F	G	H	I
1		US	Number of	Number of	Total Number	Divorces	Marriages	Divorces per	Divorces per
2		Population	Divorces	Marriages	Marriages in the	in that year per	in that year per	100	1,000
3			Occurring	Occurring	Population	1,000	1,000 Marriages	Marriages	
4	Year		in that year	in that year		Population	Population	In that year	in the Population
5	1900	75,994,575	55,751	685,284	13,885,000	0.7	9.0	8.1	4.0
6	1910	91,972,266	83,000	948,000	17,843,000	0.9	10.3	8.8	4.7
7	1920	105,710,620	171,000	1,274,000	21,333,000	1.6	12.1	13.4	8.0
8	1930	122,775,046	183,000	1,102,000	26,245,000	1.5	9.0	16.6	7.0
9	1940	131,669,275	269,000	1,566,000	30,150,000	2.0	11.9	17.2	8.9
10	1950	150,697,361	385,000	1,667,000	37,450,000	2.6	11.1	23.1	10.3
11	1960	178,464,236	393,000	1,523,000	42,200,000	2.2	8.5	25.8	9.3
12	1970	203,211,926	708,000	2,159,000	47,600,000	3.5	10.6	32.8	14.9
13	1980	222,300,000	1,189,000	2,390,000	52,300,000	5.3	10.8	49.7	22.7
14	1990	250,132,000	1,182,000	2,443,000	56,300,000	4.7	9.8	48.4	21.0
15	2000	282,339,000	1,157,589	2,329,000	60,050,000	4.1	8.2	49.7	19.3
16	2004	296,410,000	1,126,358	2,311,998	61,410,000	3.8	7.8	48.7	18.3
17	2010	308,745,538	1,111,484	2,096,000	63,799,000	3.6	6.8	53.0	17.4
18	2016	323,127,513	1,034,008	2,245,404	66,241,000	3.2	6.9	46.0	15.6

Data from: National Center Health Statistics https://www.cdc.gov/nchs/nvss/marriage-divorce.htm

1. For the year 2004 use each of the four rates in a sentence.

2. Which of these rates do you feel is the best way to report divorce rates? Explain.

3. Which rates seems to support the idea that half of all marriages end in divorce?

4. Does the conclusion that roughly half of all marriages end in divorce logically follow from these rates?

5. Consider the $x - y$ scatterplot shown below. In what year does the linear trend radically change?

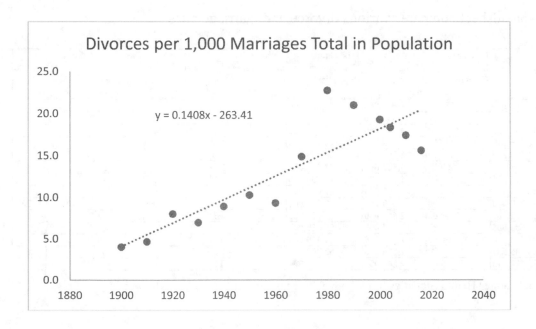

6. What could explain the divorce rate starting to fall after 1980?

7. If you knew that the number of new marriages has been declining since 1980, as more and more young couples choose to co-habitate, does that change your answer to #6?

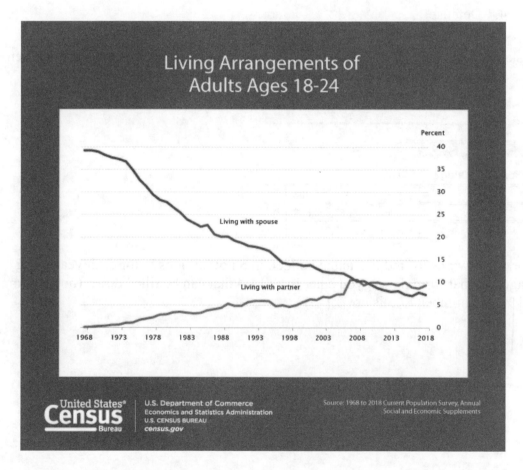

8. Interpret the slope of the line from the scatterplot in #5 as a rate.

9. Find the equation of the line going between the points (0, 22.7) and (36, 15.6) using x = years since 1980. Use your line to predict the divorce rate per 1,000 marriages total in the year 2020 (4 decimal places for the slope and 1 decimal place for the rate).

Excel Preview – 4.4 Defense

This problem asks you to look at U.S. defense spending in current dollars (not scaled to a given year) and also as a percentage of the Federal budget.

	A	B	C	D
1				
2	National Defense Spending			a.)
3	Year	(millions $)	% of Budget	Budget ($)
4	1940	1,660	17.5	$ 9,485,714,286
5	1941	6,435	47.1	$ 13,662,420,382
6	1942	25,658	73.0	$ 35,147,945,205

1. You can work backwards to find the actual budget, in $ not millions $, for each year using the percentages. What formula can you enter in cell **D4** that can be filled down (note column **C** is labeled %, the number 17.5 is not 0.175)?

2. The following scatterplot gives the best-fit equation. How can you find the exact values of the slope and intercept?

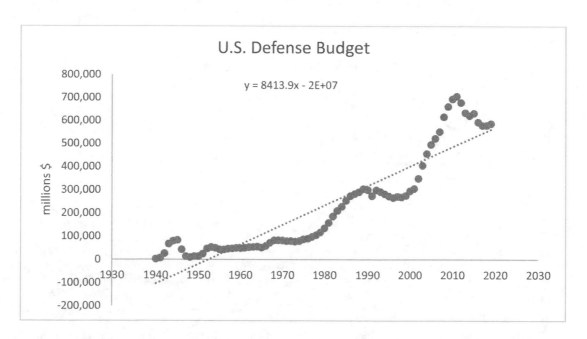

3. Interpret the actual slope and intercept in real world terms.

	Slope:	Intercept:
From Equation:	8,413.9	-20,000,000
Using SLOPE and INTERCEPT:	8,413.9	-16,426,569

4. Use the actual values to make a prediction for defense spending in 2050.

5. Now look at the scatterplot for defense spending as a percentage of the budget. Interpret the slope and intercept.

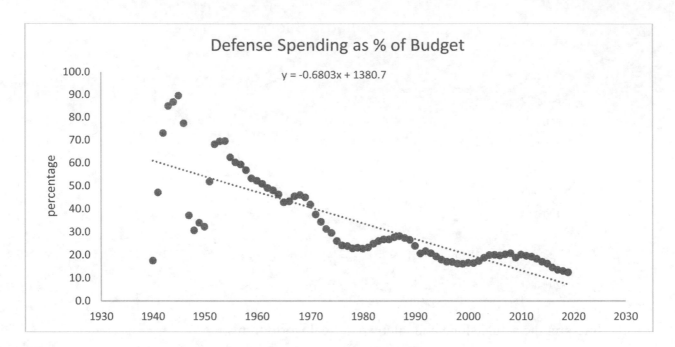

Defense Spending as % of Budget

y = -0.6803x + 1380.7

6. Predict defense spending as percentage of the budget in 2050.

7. Has defense spending been increasing or decreasing? Explain?

Section 5.4 Reflections
Monitor Your Understanding

We covered a lot of material on linear functions! Students have seen some of this material before. What was new or surprising to you?

Chapter 6 – Exponential Functions

Section 6.1 Constant Percent Change

Objective 1 – Explore Compound Interest
Objective 2 – Understand Repeated Multiplication as Exponentiation

Objective 1 – Explore Compound Interest

Summary

In Chapter 5 on linear functions we saw that the key to linearity was the slope. We can actually use the idea of the slope to define a linear function as one in which the total change in output is proportional to the total change in input. Mathematicians commonly say linear functions have constant change, referring to the constant of proportionality. The phrase "constant change" is misleading however, because there are two different types of change: total change and percent change! Linear functions have constant total change. What type of function would we get if the outputs grow by a constant percent change? We start with an example of this type of function, dealing with compound interest.

If you deposit $500 into a savings account with an **APR** of 6%. Then after 1 year you will have:
$$\$500 + 6\% \cdot \$500 = \$500 \cdot (1 + 6\%) = \$500 \cdot 1.06 = \$530$$

After two years you will have:
$$\$530 + 6\% \cdot \$530 = \$530 \cdot (1 + 6\%) = \$530 \cdot 1.06 = \$561.80$$

Notice how the interest is growing every year, 6% of $500 is $30.00, while 6% of $530 is $31.80. The extra $1.80 comes from 6% of the $30 in interest from year one. We could continue this process and see that after ten years we would have $895.42. Yes spreadsheets would do this for us very quickly, but in the next objective we will model this situation with an exponential equation.

As you read, jot down notes and questions. At the end of this section's Guided Worksheets there is a space for **Reflection** *and* **Monitoring Your Understanding**, *where you can try and answer your questions after completing the guided activities and homework.*

Notes & Questions

Guided Practice Activity #1 – Compound Interest

1. You invest $1,000 in a mutual fund that returns an average of 9% each year.

   a. Fill in the following table:

Year	Beginning	Interest	End
1	$1,000.00		
2			
3			
4			
5			
6			
7			
8			
9			
10			

   b. About how many years does it take for your money to double? Answer should be to the nearest whole number.

   c. Is your ending balance each year a linear function of the years? Why or why not?

   d. Why does the interest increase each year?

2. Can you think of an equation that would compute how much you have after 10 years?

Key Terms

Recursive formula	
Closed formula	

Summary

We have seen that if you deposit $500 into a savings account with an **APR** of 6%, then after 1 year you will have:

$$\$500 + 6\% \cdot \$500 = \$500 \cdot (1 + 6\%) = \mathbf{\$500 \cdot 1.06} = \$530$$

To make something grow by 6%, we can just multiply by the associated growth factor of 1.06! This is a huge simplification and the key to exponential growth.

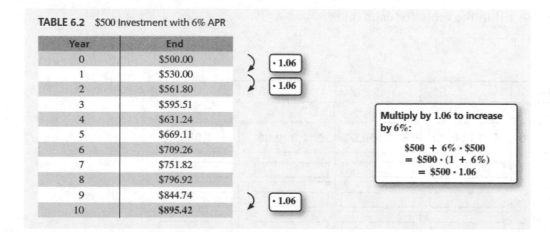

TABLE 6.2 $500 Investment with 6% APR

Year	End	
0	$500.00	· 1.06
1	$530.00	
2	$561.80	· 1.06
3	$595.51	
4	$631.24	
5	$669.11	
6	$709.26	
7	$751.82	
8	$796.92	
9	$844.74	· 1.06
10	$895.42	

Multiply by 1.06 to increase by 6%:

$$\$500 + 6\% \cdot \$500$$
$$= \$500 \cdot (1 + 6\%)$$
$$= \$500 \cdot 1.06$$

All of the values in the **End** column in **Table 6.2** can be computed from the previous year by multiplying by 1.06. Thus the value at the end of the 10th year, $895.42, comes from starting at $500 and multiplying by 1.06 ten times, i.e. $\$895.42 = \$500 \cdot 1.06^{10}$:

$$P = 500 \cdot 1.06^{t}$$

where P = principle and t = years. We are using the fact that *repeated multiplication is exponentiation*, giving us an **exponential function** (the input variable is an exponent).

As you read, jot down notes and questions. At the end of this section's Guided Worksheets there is a space for **Reflection** *and* **Monitoring Your Understanding**, *where you can try and answer your questions after completing the guided activities and homework.*

Notes & Questions

Guided Practice Activity #1 – Repeated Multiplication

1. For each of the given rates, write down the associated factor, i.e. what you multiply by to make something change by that percentage.

Rate	5%	12.8%	-6%	20%
Factor				

2. What is the formula relating the rates and the factors?

3. Use recursion to fill in the table for each rate.

Rate	5%	12.80%	-6%	20%
Year				
0	$ 1,000.00	$ 1,000.00	$ 1,000.00	$ 1,000.00
1				
2				
3				
4				
5				
6				
7				
8				
9				
10				

4. Write down the closed formula for each value in year 10:

Excel Preview – 5.1 Fund

Spreadsheets are perfect for repeated multiplication!

	A	B	C	D
1				
2				
3			Rate	Factor
4		a.)		
5				
6				
7				
8		b.)	Year	Amount
9			0	$ 1,200.00
10			1	$ 1,500.00
11			2	
12			3	
13			4	
14			5	

1. In this problem you have $1,200 invested and one year later you have $1,500.

   a. What is the percent change?

   b. What is the factor associated to this rate?

2. What recursive formula would be entered in cell **D11** that can be filled down (using cell references where possible)?

3. In the following scatterplot we can see the classic exponential growth curve and an equation of the form: $y = a \cdot e^{k \cdot x}$. What do the constants a and k equal (represent k as a percentage)?

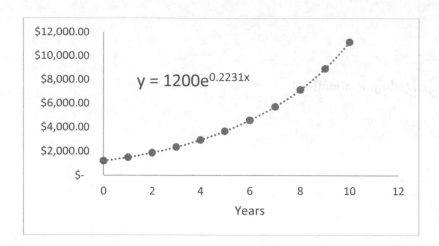

$y = 1200e^{0.2231x}$

4. What closed formula using the factor from **#1** can also model this situation?

Section 6.1 Reflections
Monitor Your Understanding

What have you learned about exponential growth that was new or different?

Objective 1 – Use Growth/Decay Rates/Factors
Objective 2 – Understand the Continuous Form

Objective 1 – Use Growth/Decay Rates/Factors

Key Terms

Exponential function	
Continuous growth/decay rate	

Summary

In the previous Excel activity, we have seen Excel display the exponential equation,

$$P = 1,200 \cdot \mathbf{1.25}^t$$

As:

$$y = 1,200 \cdot e^{0.2231 \cdot x}$$

Both of these are of the form, $= a \cdot b^x$, implying that $1.25 \sim e^{0.2231}$.

The letter e represents Euler's number (pronounced "oiler") which is an irrational number like Pi, 3.141593…, and thus can only be *approximated* by a decimal representation: 2.718282… . In the next chapter we will actually derive e and provide the precise mathematical definition (recall that Pi is the **ratio** of the circumference of any circle to its diameter). For now we need to know how to calculate with e because Excel is going to give us the equation of our exponential trendlines using Euler's number. We will use the built-in function =**EXP**(n) to calculate e^n. In particular if you type in:

$$= \textbf{EXP}(0.2231) \sim 1.2499$$

confirming the approximation above: $e^{0.2231} \sim 1.25$. Excel is rounding off the decimal which is why it is not an exact equality. All scientific calculators have an e^n button for this calculation. Note that 0.2231 is close to 0.25, and both can be interpreted as percentages: 22.31% and 25%. The **APR** is given by 25% and represents the annual **rate**, while the 22.31% represents the **continuous rate** per year and will be explained fully in the next chapter.

There are many examples of exponential growth and decay in the natural world. Populations of many species will grow exponentially for a period of time until they approach the carrying capacity of their environment, which represents the maximum population size their ecosystem can support. Human beings in particular are very adept at adapting to their environment and thus exponential growth models are very accurate for our species. Radioactive decay is an important example of exponential decay.

The fact that every exponential equation can be written in the form:

$$P = P_0 \cdot (1+r)^t,$$

illustrates our discussion at the start of the chapter regarding constant percent change. The growth/decay rate, r, is the constant rate of percent change.

As you read, jot down notes and questions. At the end of this section's Guided Worksheets there is a space for **Reflection** *and* **Monitoring Your Understanding***, where you can try and answer your questions after completing the guided activities and homework.*

Notes & Questions

Guided Practice Activity #1 – Continuous vs. Annual Rates

For each scatterplot:
- Identify the continuous rate and use it in a sentence to quantify the change of the output.
- Compute the associated rate per unit input (3 decimal places for the rate, r) and write down the equation (5 decimal places for the factor, $1 + r$):

$$P = P_0 \cdot (1+r)^t$$

1. Mutual Fund

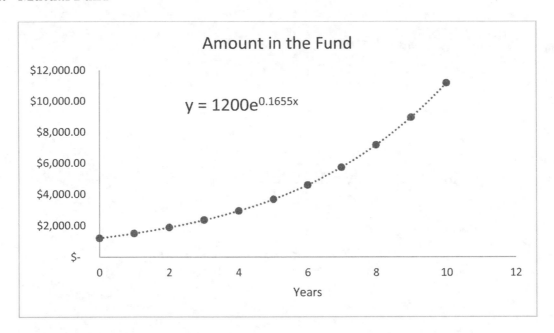

Amount in the Fund

$y = 1200e^{0.1655x}$

2. U.S. Population

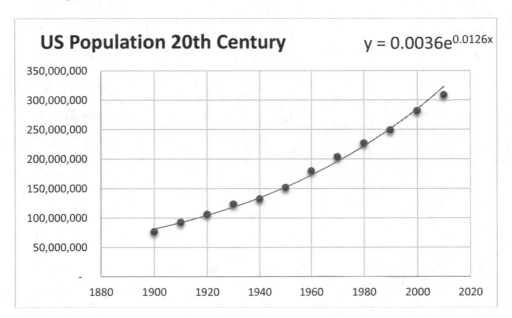

US Population 20th Century $y = 0.0036e^{0.0126x}$

3. OSHA

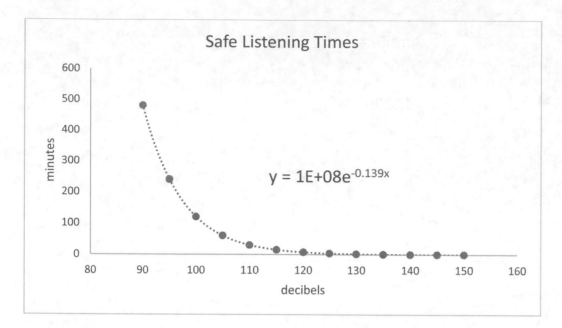

Safe Listening Times

$y = 1E{+}08e^{-0.139x}$

Objective 2 – Understand the Continuous Form

Summary

In the previous guided activity we highlighted the relationship between the continuous growth rate and the growth rate. In particular we used two important identities:

1. $e^k = 1 + r$
2. Factor = 1 + Rate (or Rate = Factor – 1)

As you read, jot down notes and questions. At the end of this section's Guided Worksheets there is a space for **Reflection** *and* **Monitoring Your Understanding***, where you can try and answer your questions after completing the guided activities and homework.*

Notes & Questions

Guided Practice Activity #1 – Carbon-14

Carbon-14 is an element that naturally builds up in pants (oops! "plants" ☺) and animals during life, and then will decay exponentially after death at a rate of 11.4% per millennia.

1. Using the equation, Factor = 1 ± Rate, determine the decay factor for Carbon-14.

2. Assuming you have 100 grams of Carbon-14 to start, fill in the following table:

Millennia (1,000's years)	Carbon-14 Remaining (g)
0	100
1	[ ]
2	[ ]
3	[ ]

3. Write down an equation for the amount of Carbon-14 remaining, C, after m millennia.

4. The graph below shows how Excel represents this equation. What is the continuous decay rate for Carbon-14?

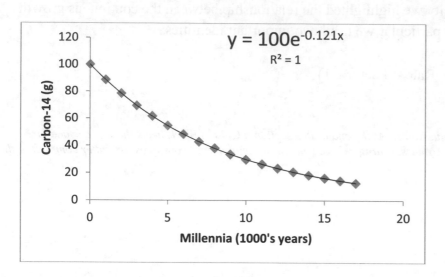

$$y = 100e^{-0.121x}$$
$$R^2 = 1$$

5. Use the equation from Excel and your e^x button on your calculator (or **EXP** function in Excel) to compute the amount of Carbon-14 left after 8,250 years.

Excel Preview – 6.2 Spill

A chemical spill has resulted in contaminated drinking water with 20 parts per billion (ppb) of the contaminant. A measurement of 20 ppb means for every 1,000,000,000 parts of a quantity of water (H_2O molecules) 20 of those parts are contaminant molecules. One day later the level of contaminant is down to 16 ppb.

	A	B
1	A chemical spill has resulted in cont	
2	of water (H_2O molecules) 20 of thos	
3		
4		
5		
6		
7		b.)
8	Day	Contaminant (ppb)
9	0	20
10	1	16
11	2	12.800
12	3	10.240
13	4	8.192

1. What is the decay rate per day and associated decay factor?

2. Assuming the decay factor is in cell **G6**, what formula can you enter in cell B11 that can be filled down?

3. What exponential equation models this situation of the form: $P = P_0 \cdot (1+r)^t$?

4. The spreadsheet shows the continuous form of the exponential equation. What is the continuous rate?

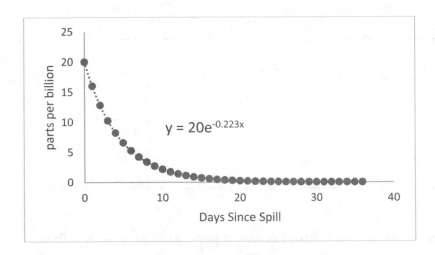

5. How much contaminant will be left after 50 days?

Section 6.2 Reflections
Monitor Your Understanding

Every exponential equation can be written in two ways. How comfortable are you working with both forms?

Section 6.3 Average Percent Change

| Objective 1 – Compare Arithmetic, Geometric, and Harmonic Means |
| Objective 2 – Compute Average Percent Change |

| Objective 1 – Compare Arithmetic, Geometric, and Harmonic Means |

Key Terms

Arithmetic mean	
Geometric mean	
Harmonic mean	

Summary

We have characterized exponential growth in terms of constant **percent change** over a fixed time period; e.g. 6% per year or -11.4% per millennium. This constant percent change is often referred to as the average percent change in much the same way as the **slope** of **linear function** is referred to as the average **total change** (or just **average change**).

If a population increases by 40 people over 5 years, then the average total change is 8 people per year. However, if a population grows by 40% over 5 years, then the average percent change is **not** 8% per year! We can easily verify this by starting with a population of 100 people, so growing by 40% (multiply by 1.40) would end with 140 people; but making 100 grow by 8% for 5 years (multiply by 1.08 five times) gives 146.9 people. To find the correct average percent change uses the geometric mean not the arithmetic mean. There is a third type of mean called the harmonic mean that is also covered here.

- **Arithmetic mean**: used to find typical "averages".
- **Geometric mean**: used to find average percent change, by taking the geometric mean of growth/decay factors.

- **Harmonic mean**: used to calculate average of certain rates, like average speeds over different parts of a trip (each part must be the same distance).

Imagine a town with a population of 900 that grows by 30% in 1 year and then 72% in the next. What is their average percentage change over the 2 years?

CAUTION! The average percentage change is NOT: $\dfrac{30\% + 72\%}{2} = 51\%$. This uses the *arithmetic mean* and does not work in this situation.

The CORRECT way to solve this is using growth factors. The town starts at 900 people and then the population is multiplied by the growth factor of 1.30 and then multiplied by 1.72:

$$900 \times 1.30 \times 1.72 = 2{,}012 = 900 \times (1+r) \times (1+r)$$

We now have two ways to proceed:

Solution Procedure #1: Use just the growth factors to solve for the average growth factor and then subtract 1 to find the average growth rate, r.

$$(1+r)^2 = 1.30 \times 1.72$$
$$1 + r = \sqrt{1.30 \times 1.72} = 1.495$$
$$r = 49.5\%$$

$$1 + r = \sqrt[n]{(1+r_1) \cdot (1+r_2) \cdots (1+r_n)}$$

Solution Procedure #2: Use the *Original* and *New* values to solve for the average growth factor and then subtract 1 to find the average growth rate, r.

$$(1+r)^2 = \frac{2{,}012}{900}$$

$$1 + r = \sqrt{\frac{2{,}012}{900}} = 1.495$$

$$1 + r = \sqrt[n]{\frac{\text{New}}{\text{Original}}}$$

$$1 + r = \sqrt[n]{\frac{\text{New}}{\text{Original}}}$$

As you read, jot down notes and questions. At the end of this section's Guided Worksheets there is a space for **Reflection** *and* **Monitoring Your Understanding***, where you can try and answer your questions after completing the guided activities and homework.*

Notes & Questions

Guided Practice Activity #1 – Mean Averages

Use the appropriate mean to solve the following:

1. You go on a trip with segments of 50 miles each and average 30 mph on the 1st segment, 50 mph on the 2nd, and 70 mph on the 3rd. What was your average speed for the entire trip?

2. A mutual fund averages 7.2%, -4.3%, and -9.1% over 3 years. What was the average rate of return over the 3 years?

3. You mix equal weights (30 kg) of lead at 11,340 kg/m^3 and gold at 19,300 kg/m^3. What is the density of the resulting alloy?

4. Apple Inc. had a net profit of $57 million in 2003 and a net profit of $14,013 million in 2010. Wow! What was the average change of Apple's net profit over this 7 year period?

    a. In $ per year?

    b. As % change per year?

Objective 2 – Compute Average Percent Change

Key Terms

Average total change	
Average percent change	

Summary

The previous objective listed two different ways to compute **average percent change** depending on the information given in the problem:

1. $1 + r = \sqrt[n]{(1 + r_1) \cdot (1 + r_2) \cdots (1 + r_n)}$

2. $1 + r = \sqrt[n]{\dfrac{\text{New}}{\text{Original}}} = \left(\dfrac{\text{New}}{\text{Original}}\right)^{1/n}$

Note that both of these formulas are computing the average **growth/decay factor,** you must subtract one to get the **rate**. We can always use the second of these equations by making up an original value and using the given growth rates to compute a new value. In general, given any two points in the plane, we can think of the output values as the original and new values. We can then compute average change, total or percent, per input unit.

As you read, jot down notes and questions. At the end of this section's Guided Worksheets there is a space for **Reflection** *and* **Monitoring Your Understanding**, *where you can try and answer your questions after completing the guided activities and homework.*

Notes & Questions

<u>**Guided Practice Activity #1**</u> – Mean Averages

Use the appropriate mean to solve the following:

1. A mutual fund increases 8% in 1 year, loses 5.2% in the next, and then is up 12.3% the following year. What is the average percentage change?

2. A town grows from 12,456 people to 32,980 in 5 years.

    a. What is the average total change per year?

    b. What is the linear equation modeling this growth?

    c. What is the average percentage change per year?

    d. What is the exponential equation modeling this growth?

The following scatterplot gives the U.S. wind energy production in billions KWhrs since 1990.

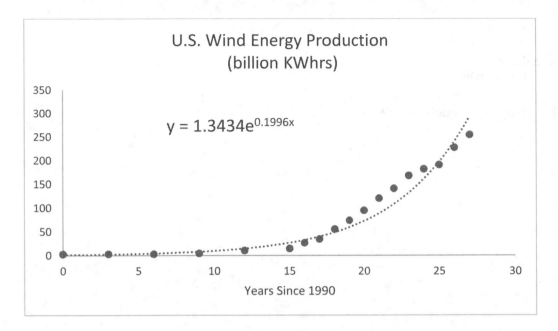

1. Use the continuous growth rate in a sentence to quantify the growth of wind energy.

2. Rewrite the given exponential equation as: $P = P_0 \cdot (1+r)^t$.

3. What is the average percent change per year?

4. Use the two points (1990, 2.8) and (2017, 254.3) and compute the average total change per year.

5. Use the two points (1990, 2.8) and (2017, 254.3) and compute the average percent change per year.

6. What is the exponential equation connecting the two points from **#5**?

Section 6.3 Reflections
Monitor Your Understanding

Can you describe when to use each mean?

Objective 1 – Forecast with LOGEST and GROWTH

Summary

Just as it does for linear functions, Excel has built-in functions that can be used to make predictions for exponential growth. The **LOGEST** function will return the growth/decay factor and the initial value of the exponential function associated with a scatterplot. For linear trendlines, we had two functions, **SLOPE** and **INTERCEPT**, to compute the two constants found in a linear function.

LOGEST is an *array* function, meaning it has two outputs: the first is the growth/decay factor and the second is the initial value. There are two options to see both outputs:

1. First enter the **LOGEST** function into a cell:
   = **LOGEST**(known_y's, known_x's)

   a. Highlight the cell with the **LOGEST** function and the cell right next to it. Press the **F2** key (**Fn + F2** on a Mac) and then press **Ctrl+Shift+Enter**. You should see both outputs next to each other.

2. Alternatively you can use the **INDEX** function to access the two outputs one at a time: =**INDEX**(**LOGEST**(known_y's, known_x's), #).

   a. If you enter 1 for the # then you will get the first output, the growth/decay factor, and if you enter 2 for the # you will get the second output, the initial value.

CAUTION! Note the **LOGEST** function returns the growth/decay factor associated with the $P = P_0 \cdot (1+r)^t$ equation.

The **GROWTH** function will calculate an output by using the exponential trendline equation. You can forecast a single value or a series of outputs:

1. Enter the **GROWTH** function for a single input:
   =**GROWTH**(known_y's, known_x's, new_x's).

2. For a series of inputs, enter these *x*-values in a column; then enter the **GROWTH** function next to the first input.

   a. For the new_x's argument, enter the range containing the series of inputs and then press **Enter**.
   b. Now highlight the cell with the **GROWTH** function and all cells below corresponding to the number of inputs. Press the **F2** key (**Fn + F2** on a Mac) and then press **Ctrl+Shift+Enter**.

As you read, jot down notes and questions. At the end of this section's Guided Worksheets there is a space for **Reflection** *and* **Monitoring Your Understanding***, where you can try and answer your questions after completing the guided activities and homework.*

Notes & Questions

Excel Preview – 6.5 Students

Our guided activity for this section will just be the Excel preview since we are dealing with Excel functions.

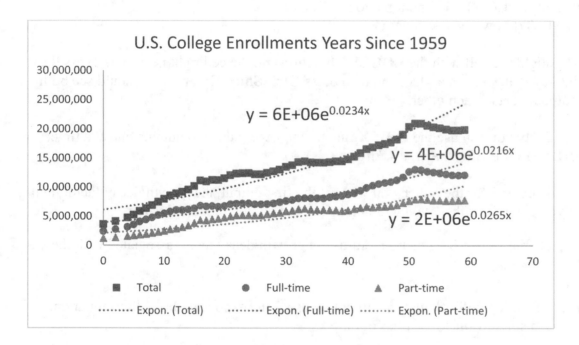

1. Which of the three student populations has been growing the fastest since 1959? What is the associated continuous growth rate?

2. The initial values are clearly rounded off in the equations. What Excel function can we use to find the exact values?

3. The inputs are in cells **A1:E59**. How would you type in the function from **#2** to find the initial value of the Part-time population?

	A	B	C	D	E
1	**a.)**	Years since 1959	Total	Full-time	Part-time
2	1959	0	3,639,847	2,421,016	1,218,831
3	1961	2	4,145,065	2,785,133	1,359,932
4	1963	4	4,779,609	3,183,833	1,595,776
5	1964	5	5,280,020	3,573,238	1,706,782
6	1965	6	5,920,864	4,095,728	1,825,136

4. How could you use the **INDEX** function to find the initial value of the Part-time population?

5. If you wanted to forecast the part-time student population in the years 2030, 2035, and 2040; what function would you use?

<u>Section 6.4 Reflections</u>
Monitor Your Understanding

Do array functions make sense to you?

Chapter 7 – Logical Arithmetic

Section 7.1 Logarithms

Objective 1 – Undo the Exponent
Objective 2 – Understand Logarithms

Objective 1 – Undo the Exponent

Summary

In Chapter 6 we encountered two different forms of the exponential equation, $y = a \cdot b^x$:

- $P = P_0 \cdot (1+r)^t$, which uses the growth/decay rate, r.

- $y = a \cdot e^{k \cdot x}$, which uses the continuous growth/decay rate, k.

Excel will display the second form as seen in the scatterplot for the U.S. Population from 1900 – 2010.

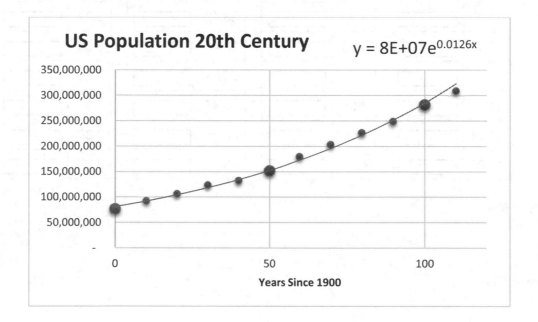

We interpreted the continuous growth rate per year of the U.S. Population to be 1.26% compared to 1.27% annual growth. In this chapter we will explore in detail the distinction between these two rates. First we look at the operation of exponentiation and its associated inverse operation, taking the logarithm. We can see the population roughly doubling every 50 years from about 75 million in 1900 to 150 million in 1950, then again doubling to 300 million in 2000. Solving for the exact doubling time requires us to solve an equation of the form:

$$2 = 1.0127^t$$

To solve for t we need to undo the exponent and get it downstairs, requiring a new arithmetic operation (logarithms), just like we need the square root operation to solve $t^2 = 9$.

As you read, jot down notes and questions. At the end of this section's Guided Worksheets there is a space for **Reflection** *and* **Monitoring Your Understanding***, where you can try and answer your questions after completing the guided activities and homework.*

Notes & Questions

Guided Practice Activity #1 – Guess and Check

Use the given tables to approximate the doubling time for the exponential equation to two decimal places. Compute the output values for the inputs, and for each rate choose the input t-value that results in an output closest to 2.

t	1.06^t		t	1.15^t		t	1.03^t		t	1.229^t
11.88			4.90			23.40			3.30	
11.89			4.91			23.41			3.31	
11.90			4.92			23.42			3.32	
11.91			4.93			23.43			3.33	
11.92			4.94			23.44			3.34	
11.93			4.95			23.45			3.35	
11.94			4.96			23.46			3.36	
11.95			4.97			23.47			3.37	

1. Rate = 6%

2. Rate = 15%

3. Rate = 3%

4. Rate = 22.9%

Objective 2 – Understand Logarithms

Key Terms

Logarithm	
Common logarithm	
Natural logarithm	
Inverse operations	

Summary

Table 7.3 compares three different common types of equations: power, linear, and exponential. To solve these equations we must use **inverse operations**; to undo a square we take the square root, to undo a product we divide, and to undo an exponent we use logarithms.

> Taking the nth root is equivalent to raising to the $1/n$.

TABLE 7.3 Comparison of Equation Types and Solution Techniques

Equation Type	Power	Linear	Exponential
Inverse operation	Roots	Reciprocals	Logarithms
Equation	$t^{1.0127} = 2$	$1.0127 \cdot t = 2$	$1.0127^t = 2$
Solution	$t = 2^{1/1.0127}$	$t = 2/1.0127$	?

The following properties are going to be useful for solving equations. The last property is the key to undoing the exponent and getting it downstairs.

General Property of Logarithms:
1. $\log_b b^k = k$
2. $\log_b b^0 = \log_b 1 = 0$
3. $\log_b b^{-k} = \log_b 1/b^k = -k$
4. $\log_b(a \cdot c) = \log_b a + \log_b c$
5. $\log_b(a/c) = \log_b a - \log_b c$
6. $\log_b a^k = k \cdot \log_b a$

As you read, jot down notes and questions. At the end of this section's Guided Worksheets there is a space for ***Reflection*** *and* ***Monitoring Your Understanding****, where you can try and answer your questions after completing the guided activities and homework.*

Notes & Questions

Guided Practice Activity #1 – Log Properties

1. Rewrite each of the following as an exponential and compute without calculator:

   a. $\log_2 8 = $ _____

   b. $\log_3 9 = $ _____

   c. $\log_{10} 10,000 = $ _____

   d. $\log 10^{-1} = $ _____

e. $\ln e^2 =$ _____

2. Use logarithms to solve for t (2 decimal places).

   a. $1.06^t = 2$

   b. $5 = 10 \cdot 0.8^t$

   c. $400 = 200 \cdot 1.15^t$

Excel Preview – 7.1 Rule of 72

The **Rule of 72** estimates how long it takes for something to double given a growth rate: given $x\%$ the doubling time $= \dfrac{72}{x}$. For example: 6% growth annually means doubling time $= \dfrac{72}{6}$ $= 12$ years. Note this is an estimate and that some people use 70 instead of 72. Guess what they call their rule!

	A	B	C	D
1				
2		**a.)**	**b.)**	**c.)**
3	APR (%)	Doubling time (years) using Rule of 72	Future Value of $1,000	Error
4	1	72.0	$ 2,047.10	$ 47.10
5	2	36.0	$ 2,039.89	$ 39.89
6	3	24.0	$ 2,032.79	$ 32.79
7	4	18.0	$ 2,025.82	$ 25.82
8	5	14.4	$ 2,018.95	$ 18.95

1. What formula is in cell **B4** that can be filled down?

2. In cell **C4** what formula can you enter that can be filled down to compute the value of $1,000 if it grows by the APR in column **A** for the number of years in column **B**?

3. The homework asks you to think about if the doubling times in column B are an exponential function of the APR. What do you think?

4. To check for exponential decay remember that when we increase the input by 1 unit then the output should be multiplied by the same decay factor every time. Divide each value in column B by the previous to compute this factor. Do you get a constant number?

APR (%)	Dbling Time	Factor
1	72.0	
2	36.0	☐
3	24.0	☐
4	18.0	☐
5	14.4	☐
6	12.0	☐
7	10.3	☐

Section 7.1 Reflections
Monitor Your Understanding

Are you able to explain what logarithms are to someone?

Section 7.2 Doubling Times and Half-lives

Objective 1 – Use Properties of Logarithms

Key Terms

Doubling time	
Half-life	

Summary

We saw that when finding the time it takes for the U.S. Population to double using the equation, $P = P_0 \cdot (1.0127)^t$, it did not matter what the initial population was. For exponential growth the doubling time depends only on the growth rate. When computing the half-life for exponential decay the rate cannot be less than -100%. It is impossible to lose more than 100% of what you have.

As you read, jot down notes and questions. At the end of this section's Guided Worksheets there is a space for **Reflection** *and* **Monitoring Your Understanding**, *where you can try and answer your questions after completing the guided activities and homework.*

Notes & Questions

Guided Practice Activity #1 – Doubling

1. You win a radio show contest and have 1 minute to choose between receiving $10,000 or you can get the sum of the following payouts for 30 days:

Day 1	**1 cent**
Day 2	2 cents
Day 3	4 cents
Day 4	8 cents
… continue doubling	
Day 30	
SUM:	???

   a. How much money do you get on day 5?

   b. How much money do you get on Day 30?

   c. How much do you get in total? Hint: Use a spreadsheet.

2. **The Parable of the Bacteria in a Bottle**[1]: Once upon a time, at precisely 11:00 pm, a single bacterium was placed into a nutrient-filled bottle in a laboratory. The bacterium immediately began gobbling up nutrients, and after just one minute — making the time 11:01 — it had grown so much that it divided into two bacteria. These two ate until, one minute later, they each divided into two bacteria, so that there were a total of four bacteria in the bottle at 11:02. The four bacteria grew and divided into a total of eight bacteria at 11:03, sixteen bacteria at 11:04, and so on. All seemed fine, and the bacteria kept on eating happily and doubling their number every minute, until the "midnight catastrophe." The catastrophe was this: At the stroke of 12:00 midnight, the bottle became completely full of bacteria, with no nutrients remaining — which meant that every single one of the bacteria was suddenly doomed to death. We now turn to our questions, as we seek to draw lessons from the tragic demise of the bacterial colony.

   a. The catastrophe occurred because the bottle became completely full at 12:00 midnight. When was the bottle *half*-full?

---

[1] Adapted from Jeffrey Bennett's excellent book, *Math for Life*.

b. You are a mathematically sophisticated bacterium, and at 11:56 you recognize the impending disaster. You immediately jump on your soapbox and warn that unless your fellow bacteria slow their growth dramatically, the end is just 4 minutes away. Will anyone believe you?

c. It's 11:59 and your fellow bacteria are finally taking your warnings seriously. Hoping to avert their impending doom, they quickly start a space program, sending little bacterial spaceships out into the lab in search of new bottles. To their relief, they discover that the lab has four more bottles that are filled with nutrients but have no one living in them. They immediately commence a mass migration through which they successfully redistribute the population evenly among the four bottles just in time (at midnight all 4 bottles are one-quarter full) to prevent the midnight catastrophe. How much more time do the additional bottles buy for their civilization?

d. Because the four extra bottles bought so little time, the bacteria keep searching out more and more bottles. Is there any hope that additional discoveries will allow the colony to continue its rapid growth? Assume each bacterium occupies a volume of 10^{-21} cubic meters and the surface area of the Earth is 510 million square km. At 1:00 AM how deep are the bacteria if they are evenly spread over the surface of the earth?

Guided Practice Activity #2 – Halving

The total amount of contaminant in a pond decreases by 25% each day.

1. Assuming you started with 100 ppm in the water, how long would you estimate until it is half gone?

2. Fill in the following table:

Days	0	1	2	3	...	t
Amount (ppm)	☐	☐	☐	☐		☐

3. Write down the equation for solving when the amount will be 50, exactly one-half of the original.

4. Solve for the "half-life" using logarithms.

5. Solve for daily continuous decay rate using logarithms.

6. Fun Balderdash Question ☺, make up a believable definition of "logarithm".

> **A logarithm** is ...

Excel Preview – 7.4 Password

This problem gives the times to crack passwords of varying lengths and also based on the number of possible symbols you are choosing from.

	A	B	C	D	E
1	The time a computer takes to find a password (by randomly trying all possibiliti				
2	computers using brute force algorithms in 2014. *Alphabet* means passwords c				
3					
4					
5					
6	Time to crack password of length:				
7	Number of	**Alphabet (26)**		**Any Symbol (94)**	
8	symbols	**abc...**		**AaBb123!@#...**	
9	4	4.57	seconds	13	minutes
10	5	1.98	minutes	20.4	hours
11	6	51.5	minutes	2.63	months
12	7	22.3	hours	20.6	years
13	8	24.2	days	1.93	millenia
14	9	1.72	years	182	millenia
15	10	44.88	years	17079	millenia

1. Use the number 2.63 from the table above in a meaningful sentence.

2. Convert the first 6 numbers to hours.

Hours to crack password of length:

Number of symbols	Alphabet	Any Symbol
4		
5		
6		

3. The following scatterplot gives exponential models for the times. Interpret the continuous growth rate for passwords consisting of alphabet symbols.

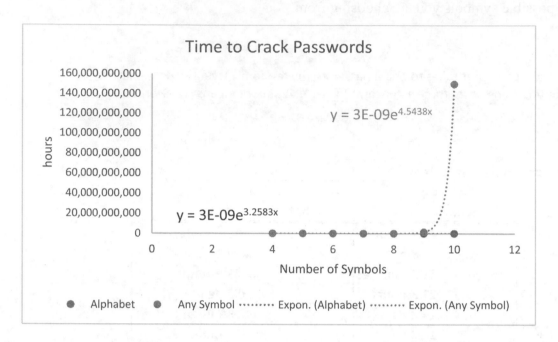

4. Why can we see essentially only one data value? And which value is it?

5. Any brilliant ideas for how to plot this data so we can see all the data values?

Section 7.2 Reflections
Monitor Your Understanding

Are there other functions besides exponential that have doubling times and half-lives?

Section 7.3 Annual Percentage Yield

Objective 1 – Understand Compounding Periodically
Objective 2 – Compute the Yield (APY)

Objective 1 – Understand Compounding Periodically

Key Terms

Periodic compounding form	

Summary

We have defined exponential growth in terms of a constant percent change using the equation, $P = P_0 \cdot (1+r)^t$. This equation works for periodic rates as well. Recall that if we have a 6% APR compounded monthly, then the periodic or monthly rate is $\dfrac{6\%}{12} = 0.5\%$. Assuming you invest $1,000 with a 6% APR compounded monthly, we can write the equation for the value of your investment as a **function** of months:

$$P = 1,000 \cdot (1+0.5\%)^m$$

Where m = months and P = the principal or value of our investment after m months. Recalling our work in Chapter 3, the **ratio**, 1 year : 12 months, yields the equation $m = 12 \cdot t$, which we can substitute into the equation above:

$$P = 1,000 \cdot \left(1 + \frac{6\%}{12}\right)^{12 \cdot t}$$

This leads us to the **periodic compounding form** of an exponential equation:

$$P = P_0 \cdot \left(1 + \frac{APR}{n}\right)^{n \cdot t}$$

Where P_0 = the initial value and n = the number of periods in one year.

As you read, jot down notes and questions. At the end of this section's Guided Worksheets there is a space for **Reflection** and **Monitoring Your Understanding**, where you can try and answer your questions after completing the guided activities and homework.

Notes & Questions

Guided Practice Activity #1 – Periodic Compounding

1. Write down the exponential equation for the amount of money, P, after m months at 8% compounded monthly starting with $1,000.

2. Note the relationship between Months and Years by filling in the following table:

Months (m)	12	48	6	3	$m = 12 \cdot t$
Years (t)					

3. Rewrite your equation from #1 using years, t, as the input.

4. Check that $P(12)$, $m = 12$, from question #1 equals $P(1)$, $t = 1$, from question #3.

5. For each of the following write down the periodic compounding form and the amount after 5 years:

    a. $5,000 at 4.5% compounded daily

    b. $20,000 at 9.2% compounded quarterly

c. $1 million at 7.4% compounded annually

d. $1 million at 7.4% compounded monthly

Objective 2 – Compute the Yield (APY)

Key Terms

APY	
Effective interest rate	

Summary

The **periodic compounding form** of an exponential equation is a bit messy, but will allow us to compare different investments as the input variable will always be years. Also given an APR and period we can simply "fill in the blanks" in the general form. Given an investment of $1,000 at 6% compounded monthly, we can represent this using the equation:

$$P = 1,000 \cdot \left[\left(1 + \frac{6\%}{12} \right)^{12} \right]^{t}$$

Notice everything in the square brackets is a number, typing this all into a calculator we can simplify:

$$P = 1,000 \cdot (\mathbf{1.061678})^{t}$$

That looks much more manageable and notice it is of the form, $= 1,000 \cdot (1 + r)^{t}$, implying the **growth rate** $r = 6.1678\%$. This represents what an **APR** of 6% compounded monthly will yield at the end of the year. If we substitute $t = 1$ we get $1,061.68, i.e. we get 6.1678% interest. This new rate is called the annual percentage yield, or APY.

As you read, jot down notes and questions. At the end of this section's Guided Worksheets there is a space for **Reflection** *and* **Monitoring Your Understanding**, *where you can try and answer your questions after completing the guided activities and homework.*

Notes & Questions

Guided Practice Activity #1 – APY

Fill in the table below following the prompts:

1. First compute the periodic rates.

2. Next enter the names of the compounding periods (monthly, weekly, etc.)

3. Compute the value of a $1,000 investment after 1 period for each periodic rate. Not 1 year but just 1 period (1 month, 1 day etc.).

4. Compute the value of a $1,000 investment after 1 year for each periodic rate.

5. Finally compute the APY for each periodic rate.

APR	8%	8%	8%	8%	8%	8%	8%
Nper in 1 Year	1	2	4	12	26	52	365
Periodic Rate							
Period Names							
Period 0	$1,000.00	$1,000.00	$1,000.00	$1,000.00	$1,000.00	$1,000.00	$1,000.00
Period 1							

Year 0	$1,000.00	$1,000.00	$1,000.00	$1,000.00	$1,000.00	$1,000.00	$1,000.00
Year 1							
APY							

This spreadsheet considers an investment of $10,000 with a 6.5% APR for different compounding periods.

	A	B	C	D	E
1	Principal:	$ 10,000.00		Given an investment of $10,000 with a 6.5% APR compu	
2	APR	6.50%			
3					
4		a.)	b.)	c.)	d.)
5	Compounding Periods	Periods in 1 Year	Yield	Doubling Time (years)	Amount (after 15 years)
6	Annually				
7	Monthly				
8	Daily				
9	Hourly				
10	Secondly				
11	Continuously				
12					

1. Assuming the periods in 1 year are filled in correctly, what formula can you enter in cell **C6** that can be filled down?

2. What formula can be filled down for cell **D6** using the yields from column **C**?

3. What formula can be filled down for cell **E6**?

4. The "Continuously" row is different from the others. How many periods would there be if it was compounding continuously?

5. Any idea how to calculate the yield, doubling time and amount for this row?

Can you describe the difference between APR and APY?

Section 7.4 Continuous Growth and Euler's Number: *e*

Objective 1 – Explore Compounding Continuously

Key Terms

Euler's number	

Summary

The **periodic compounding form** of an exponential equation is what will lead us to *e* and the concept of the **continuous growth/decay rate**. The idea is simple enough; if we can compound our interest twelve times a year, i.e. every month, then we can compound more often: every day, every minute, every second, every nanosecond, etc. As *n*, the number of periods in a year, gets larger, the associated period of time gets smaller. Eventually the period of time will be so short that it will be indistinguishable from continuous compounding. Think of it like a movie which is nothing more than a collection of pictures flashed before us so quickly our eyes cannot distinguish the "flip rate" and it appears continuous.

In the chapter we analyze what happens to the periodic compounding equation with 6% APR as the number of compounding periods in one year goes to infinity.

$$P = P_0 \cdot \left[\left(1 + \frac{1}{N} \right)^N \right]^{6\% \cdot t}$$

As *N* goes to infinity the expression in the square bracket goes to *e* giving us the equation:

$$P = P_0 \cdot \left[e \right]^{6\% \cdot t}$$

This is exactly the continuous growth form of an exponential equation, $y = a \cdot e^{k \cdot x}$, that we saw in Chapter 6. Now we know that the word "continuous" is referring to getting interest compounded continuously throughout the year.

As you read, jot down notes and questions. At the end of this section's Guided Worksheets there is a space for **Reflection** *and* **Monitoring Your Understanding**, *where you can try and answer your questions after completing the guided activities and homework.*

Notes & Questions

Guided Practice Activity #1 – The Big Enchilada (wrapping it all together)

1. In the equation:

$$P = 1,000 \cdot \left(1 + \frac{7.2\%}{26}\right)^{26 \cdot t}$$

a. What is the APR?

b. What is the period?

c. What is the periodic rate as a percentage to 3 places?

d. What is the APY as a percentage to 3 places?

e. What is the annual growth factor (to 5 decimal places)?

f. What is the associated continuous rate as a percentage to 3 places (that has the same APY)?

g. What is the growth factor per decade (to 2 decimal places)?

h. To find the doubling time what is set equal to $2,000?

i. What is the doubling time (to 2 decimal places)?

2. Fill in the following table:

Annual Growth/Decay Rate	Doubling Time/Half-Life		Growth/Decay factor per year	Continuous rate per year	Growth/Decay factor per decade	Growth /Decay rate per decade
r	$\dfrac{\log 2}{\log 1+r}$	$\dfrac{\log 1/2}{\log 1+r}$	$1+r$	$\ln(1+r)$	$(1+r)^{10}$	$(1+r)^{10}-1$
30%	[ ]		1.30	[ ]	13.79	[ ]
[ ]	9.0 years		1.08	7.70%	[ ]	116%
[ ]	[ ]		[ ]	[ ]	0.197	[ ]
-25%	[ ]		[ ]	[ ]	0.0563	[ ]
[ ]	[ ]		1.043	4.21%	[ ]	[ ]

Excel Preview – 7.2 Table

The following table asks you to use formulas in Excel to complete. Assume all cells will be formatted correctly.

	A	B	C	D	E	F	G
1							
2							
3		Annual Growth/Decay Rate	Doubling Time/ Half-Life	Growth/Decay Factor per Year	Continuous Rate per Year	Growth/Decay Factor per Decade	Growth /Decay Rate per Decade
4	a.)	17.2%		1.172		4.890	
5	b.)		2.790	0.780	-24.8%		
6	c.)		6.785			0.360	
7	d.)				3.2%		-100.0%
8							

1. Recall that the growth factor per year is the key. What formula would you type into cell **D6** to compute the factor for a rate with a growth factor per decade of 0.36?

2. What formula would you type into cell **D7** to compute the factor for an annual rate with a continuous rate of 3.2%?

3. What formula can be typed into cell **B5** that can be filled down?

4. What formula can be typed into cell **C4**?

5. What formula can be typed into cell **E4** that can be filled down?

6. What formula can be typed into cell **F5** that can be filled down?

7. What formula can be typed into cell **G4** that can be filled down?

Section 7.4 Reflections
Monitor Your Understanding

Can you describe what it means to compound continuously?

Objective 1 – Understand Log Scales

Key Terms

Logarithmic scale	
Linear scale	
Semi-log plot	
Log-log plot	
Power law distribution	
Power law equation	

Summary

There is one other prevalent use of **logarithms**, related to graphing: the **log scale**. This occurs in the Richter scale for measuring earthquake magnitudes and the decibel scale for measuring "loudness" of sounds. These examples are explored in the homework. A log scale is effective when our data varies over a huge range of numbers, like the planetary distances from the sun in our solar system as seen in the following scatterplot.

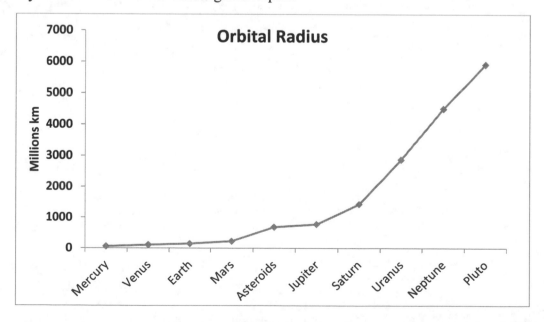

Note the linear scale on the *y*-axis. It increases in even increments of 1,000. The graph is difficult to read because the first four radii values are so much smaller than the later radii, so the first four look almost like they are zero, squished down on the *x*-axis. To deal with this, scientists create what is called a log plot using a log scale.

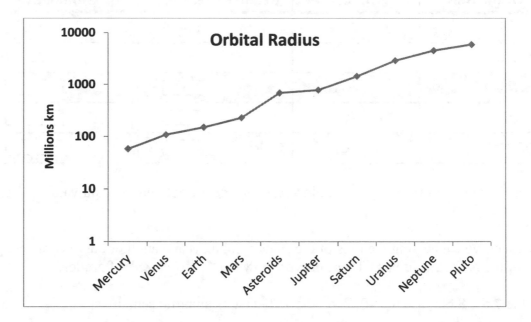

Note the logarithmic scale on the *y*-axis. It increases by powers of 10. You can now see that Earth and Venus have orbital radii closer to 100 than 10. Note that the **logarithmic scale** turned what looked like an exponential curve into a linear curve. A logarithmic scale will always turn an exponential graph into a linear graph; as explained in the chapter.

*As you read, jot down notes and questions. At the end of this section's Guided Worksheets there is a space for **Reflection** and **Monitoring Your Understanding**, where you can try and answer your questions after completing the guided activities and homework.*

Notes & Questions

Guided Practice Activity #1 – Log Scales and Cumulative Frequency Distributions

1. Fill in the following table for a hypothetical data set:

Grade Bins	Counts	Relative Frequency	Cumulative
5...6.5	20	10.0%	10.0%
6.5...7.5	50	25.0%	35.0%
7.5...8	80	☐	☐
8...9.5	40	☐	☐
9.5...10	10	☐	☐

2. For each column in the histogram below write inside the column the data value associated to it.

3. Draw a curve connecting the tops of the shaded columns and another curve connecting the tops of the hashed columns. These curves show the "*distribution*" of grades.

4. For the **7.5 .. 8** bin, use the 40% and the 75% in a meaningful sentence.

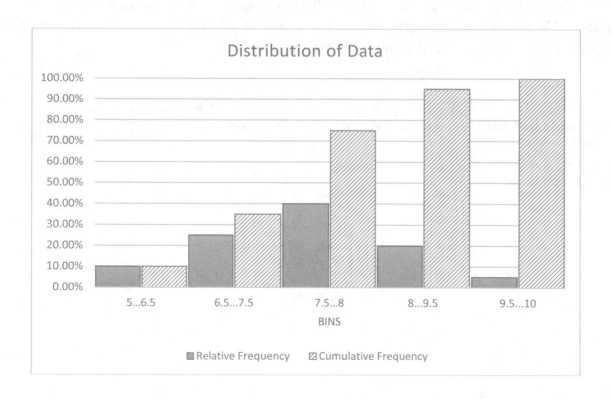

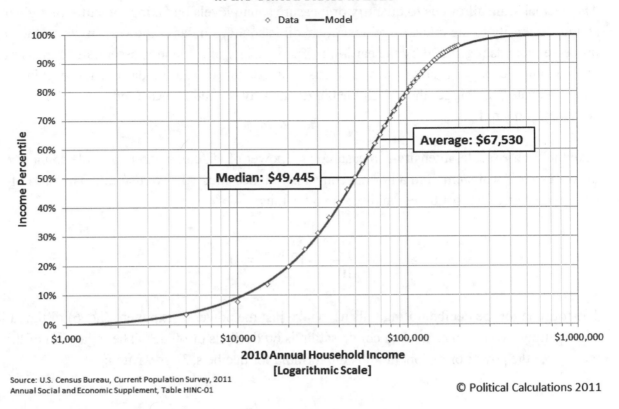

Cumulative Distribution of Total Money Income Earned by Households in the United States in 2010

◇ Data ——Model

Income Percentile (y-axis): 0%, 10%, 20%, 30%, 40%, 50%, 60%, 70%, 80%, 90%, 100%

Median: $49,445

Average: $67,530

2010 Annual Household Income
[Logarithmic Scale]

x-axis: $1,000 $10,000 $100,000 $1,000,000

Source: U.S. Census Bureau, Current Population Survey, 2011
Annual Social and Economic Supplement, Table HINC-01

© Political Calculations 2011

5. This graph shows the household income distribution in the United States in 2010. What percentage of households make less than $100,000?

6. What percentage of households make between $50,000 and $70,000?

7. How much do you have to make to be in the top 10% of all household incomes?

8. What is strange about the x-axis scale?

9. Why is this called a log scale?

The decibel scale allows us to quantify or measure sound levels by taking the **ratio** of a given sound's intensity to the intensity of a barely audible sound, similar to the way in which the brain processes sounds registered by the eardrum. The intensity of a sound is essentially a measure of the amount of energy the sound wave imparts on a flat surface at a standard distance (think waves in water, the bigger the splash the bigger the waves), and is measured in watts per centimeter squared.

Like the Richter scale, intensity is measured as a power of 10, e.g. a barely audible sound measures 10^{-16} watts/cm^2 and a normal conversation measures 10^{-10} watts/cm^2. The ratio of a normal conversation to a barely audible sound is thus:

$$\frac{10^{-10}}{10^{-16}} = 10^{-10-16} = 10^6$$

The formula for the decibel rating tells us to take just the power or exponent (i.e. 6) of the ratio, and multiply by 10, thus a normal conversation is 60 decibels or 60 dB. The logarithm of the ratio gives the power or exponent, thus the formula for decibels, D, is written:

$$D = 10 \cdot \log\left(\frac{I}{I_0}\right)$$

Where I^0 = intensity of sound you are measuring, and I^0 = the intensity of a barely audible sound, 10^{-16} watts/cm^2. Notice that the decibel scale is a log scale like the Richter scale, but we always divide by 10^{-16} watts/cm^2.

1. Compute the decibel rating of a chain saw at 10^{-6} watts/cm^2 and a dance club at 10^{-5} watts/cm^2.

2. What is the decibel rating of a barely audible sound at 10^{-16} watts/cm^2?

3. Working backwards is always hard. In this part we are given a decibel rating and want to recover the intensity. Note in the example above that 60 db comes from the ratio $\dfrac{10^{-10}}{10^{-16}} = 10^6$, thus the 6 tells us the intensity associated to 60 db must have an exponent 6 more than −16, in other words −10. A decibel rating of 110 db implies the intensity must have an exponent that is 11 more than −16, or −5. Thus the intensity of a 110 db sound is 10^{-5} watts per square centimeter.

   a. A mosquito buzzing has a decibel rating of 40 dB while the telephone dial tone is 80 dB. Compute the intensity in watts per square centimeter for both of these sounds and format as scientific notation.

   b. Is it correct to say the dial tone is twice as loud as the mosquito? Explain!

4. Graphing the intensities of sounds for given decibels results in the following scatterplot. Explain why a log plot is needed.

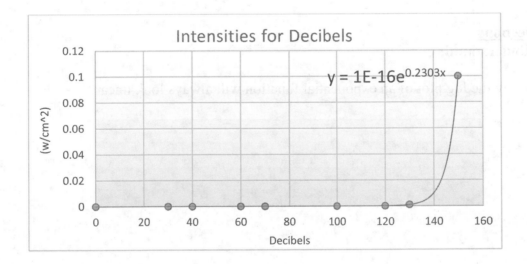

5. Switching to a log plot gives the following scatterplot. The points all look like they are below the *x*-axis. Explain why this is not so!

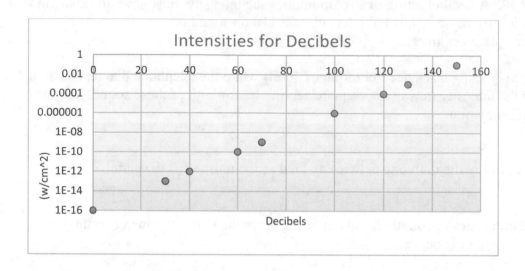

Section 7.5 Reflections
Monitor Your Understanding

Can you explain why the log plot of an exponential function will always look linear?

Chapter 8 – Curve Fitting and Correlation

Section 8.1 Correlation, Causation, and Confounding factors

Objective 1 – Understand Correlation
Objective 2 – Explore Bivariate Data

Objective 1 – Understand Correlation

Summary

Spread of Wal-Mart Supercenters Linked to Obesity read the March 9, 2011 Harvard Business Review Daily Stat headline. The caption went on to state:

> "Research suggests that one additional Wal-Mart Supercenter per 100,000 residents increases individuals' probability of being obese by 2.3 **percentage points**. The researchers, published in the *Journal of Urban Economics*, say their study implies the proliferation of Wal-Mart Supercenters, which offer food at highly discounted prices, explains 10.5% of the U.S. rise in obesity since the late 1980's."

Please read the caption carefully again and try to understand what the researchers are supposedly claiming is the link between supercenters and obesity. This chapter will go over the mathematics behind such statements. The caption is implying a **causal** link between supercenters and obesity. Namely that access to mass quantities of inexpensive unhealthy food at supercenters contributes to unhealthy weight gain. Supercenters are part of the *cause* people become obese; is the claim as stated in the caption. This claim may seem obvious, but having objective data takes us from the realm of speculation to science.

The researchers most definitely can point to a **correlation**; people who live in an area with a higher density of supercenters (relative to the population) are heavier than people who live in a less supercenter-dense area. In fact, the researchers are saying that if a city has two supercenters per 100,000 people and 6% of these people are obese, then another city with three supercenters per 100,000 people would have 8.3% of the population obese. Correlation, however, does not imply **causation**, which does not disprove the researchers' claim of causation; but does mean that one must carefully read the research article which details the study and methods for taking confounding factors into account. **Confounding factors** are other possible factors that could cause the observed behavior. For instance, maybe there are more supercenters built in poorer areas and obesity is higher in poorer regions in general.

*As you read, jot down notes and questions. At the end of this section's Guided Worksheets there is a space for **Reflection** and **Monitoring Your Understanding**, where you can try and answer your questions after completing the guided activities and homework.*

Notes & Questions

Guided Practice Activity #1 – Correlation or Causation?

For each of the following situations write down whether you think there is a true causal relationship and if causal list a possible causal agent. Then state whether it is a positive relationship (if one increases the other also increases) or negative relationship (if one increases then the other one decreases).

1. Shootings per day in a city vs. Daily temperature in the city
   https://www.nytimes.com/2018/09/21/upshot/a-rise-in-murder-lets-talk-about-the-weather.html

2. Divorce Rate in Maine vs. Per capita consumption of margarine
   http://tylervigen.com/spurious-correlations

3. Annual greenhouse gas (GHG) emissions vs. Arctic sea ice extent
   https://www.epa.gov/climate-indicators/climate-change-indicators-arctic-sea-ice

4. Density of yoga studios vs. Obesity rate
   https://www.sciencedirect.com/science/article/pii/S0965229913000964

5. Sale of winter boots at L.L. Bean in Freeport, ME vs. Sale of ice cream cones at same store

6. Number of math Ph. D's awarded by country vs. Average annual temperature

7. Average annual temperature by county vs. Average PSAT scores by county
   https://scholar.harvard.edu/files/joshuagoodman/files/w24639.pdf

Objective 2 – Explore Bivariate Data

Key Terms

Univariate data	
Bivariate data	
Correlation	
Causation	
Confounding factors	

Summary

Mathematics classes tend to deal with data pairs (x, y) where y is a **function** of x and the data follow a perfect pattern, e.g. $y = 3 \cdot x + 7$. In statistics, we have been looking at a single variable, e.g. $X =$ women's heights, and then describing the variability of this **univariate data** according to the shape of the **distribution**, and measures of center and spread. In this chapter we will look at **bivariate data**, e.g. $X =$ Walmart supercenter density and $Y =$ obesity rates for regions. The intent with bivariate data analysis is to try and explain causal relationships using scatterplots and measures of correlation.

Understanding how to read scatterplots is thus a very important skill, and many students make the mistake of interpreting all scatterplots as being time series (have time on the x-axis).

As you read, jot down notes and questions. At the end of this section's Guided Worksheets there is a space for **Reflection** *and* **Monitoring Your Understanding***, where you can try and answer your questions after completing the guided activities and homework.*

Notes & Questions

Guided Practice Activity #1 – Scatterplot Savvy

The following data are shown in a table (first five years) and also as two time series line graphs.

Year	Min Wage (2012$)	% of hourly paid workers
1979	$8.97	13.4
1980	$8.46	15.1
1981	$8.29	15.1
1982	$7.82	12.8
1983	$7.59	12.2

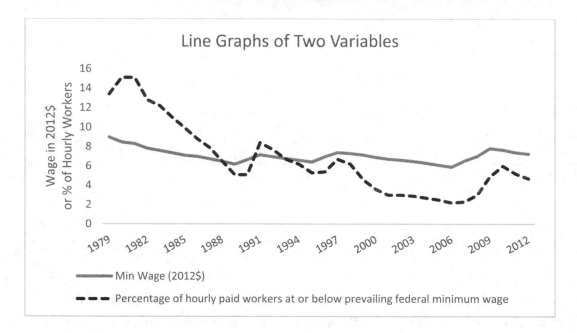

1. What can you say about the percentage of hourly paid workers at or below prevailing federal minimum wage over time?

2. What can you say about the prevailing federal minimum wage (2012$) over time?

3. Now consider the bivariate scatterplot. What do we gain with the scatterplot and what information do we lose?

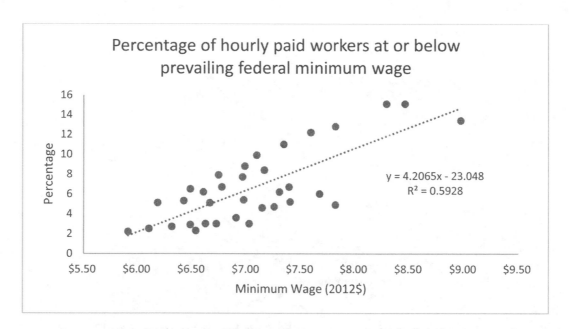

4. Circle the 1979 and 1983 points (from table above) on the scatterplot.

5. Does there seem to be a correlation between the two variables?

6. Is there a causal relationship?

7. Interpret the slope of the line in real world terms.

8. Interpret the *y*-intercept of the line in real world terms.

Excel Preview – 8.6 Humidity

Most people are aware that humidity makes it feel hotter than the actual temperature. The scatterplot represents the heat index (apparent temperature) versus the relative humidity and actual temperature.

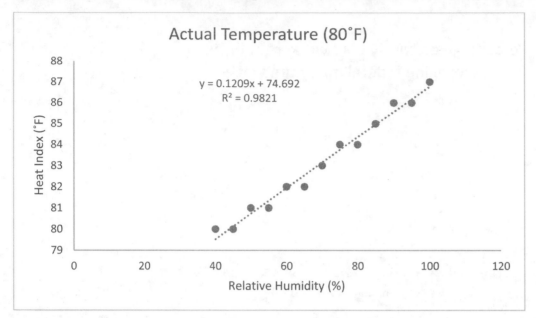

1. What is the causal mechanism at work here? Why does humidity make it feel hotter?

2. If the relative humidity is 75% how hot (1 decimal place) does it feel on an 80 degree day?

3. How much does the apparent temperature increase for every 10 percentage point increase in relative humidity? (1 decimal place)

4. Interpret the y-intercept in real world terms.

Section 8.1 Reflections
Monitor Your Understanding

Can you describe the difference between correlation and causation?

Section 8.2 Best-Fit, Least-Squares, and Regression Lines

Objective 1 – Understand Sum of Squared Errors
Objective 2 – Explore Perfect Linear Fit

Objective 1 – Understand Sum of Squared Errors

Key Terms

Least-squares line	

Summary

The scatterplot below shows a linear trendline fit to the scatterplot of oil and gas prices. Note how linear this bivariate data looks. As oil prices rise, the price of a gallon of gasoline rises in sync and appears to model a true causal relationship; rising oil prices cause the price of gas to also rise. Why does Excel choose this line as the best model for the price of gasoline as a linear function of the price of oil? If we wish to make predictions about the future price of gas assuming oil prices increase, why should we use this line? To answer these questions we will look at the criteria that statisticians have agreed upon for choosing the best fit linear model. Using the term *error* or *deviation* for the vertical distance each data point is from the line, we seek a best fit line that minimizes the sum of the squared errors. Thus the best fit line is also called the least squares lines.

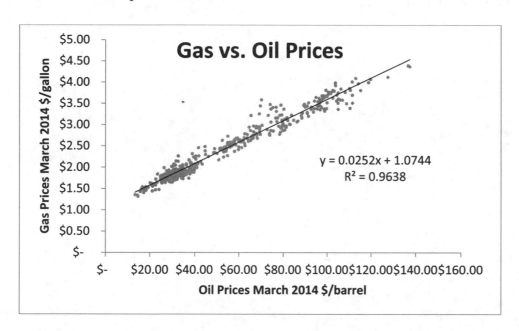

CAUTION! We minimize the sum of the **squared** errors not the sum of the errors. The errors will be both positive and negative (above and below the line), which will effectively cancel when summed. Thus we use squared errors to make everything non-negative. Squaring has the added benefit of "rewarding" small errors and "penalizing" large errors.

As you read, jot down notes and questions. At the end of this section's Guided Worksheets there is a space for
Reflection *and* ***Monitoring Your Understanding****, where you can try and answer your questions after completing the guided activities and homework.*

Notes & Questions

Guided Practice Activity #1 – Summing the Squares

Since World War II, weapons-grade plutonium for the United States nuclear weapons has been produced at the Hanford facility in southeastern Washington State, bordering on the Columbia River. Downriver from the facility the Columbia flows into northern Oregon, the Portland metropolis, and into the Pacific Ocean. There has been concern for many years about leakage from radioactive waste sites at Hanford and potential health risks related to it. Currently the huge facility is undergoing a massive and costly clean-up which is scheduled to be completed in year 2030. One of the very earliest scientific studies of the problem was conducted in the early 1960's (Fadeley, R. C., "Oregon Malignancy Pattern Physiographically Related to Hanford, Washington, Radioisotope Storage", *Journal of Environmental Health*, 1965). In the study a measure called "the index of exposure" was developed to attempt to quantify the degree of radioactive contamination at any specific location. The investigators computed this index for nine Oregon counties along the Columbia River. They then sought to relate the index to the cancer mortality rate for each county.

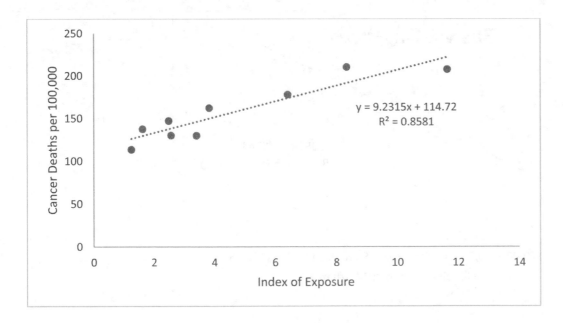

County	Index of Exposure	Cancer Deaths per 100,000	Best-fit y-values	Errors	Errors squared
Umatilla	2.49	147.1			
Morrow	2.57	130.1			
Gilliam	3.41	129.9			
Sherman	1.25	113.5			
Wasco	1.62	137.5			
Hood River	3.83	162.3			
Portland	11.64	207.5			
Columbia	6.41	177.9			
Clatsop	8.34	210.3			

1. Identify the point on the scatterplot corresponding to the first county, Umatilla.

2. Compute the y-values for the best-fit line by substituting x-values into the equation and filling in the table (1 decimal place).

3. Compute the errors by subtracting the y-values, if the data point is above the line the error should be positive (1 decimal place).

4. Compute the errors squared (1 decimal place).

5. Compute the sum of the squared errors.

6. Why is the best-fit line in the scatterplot called the "least-squares" line?

7. Compute the sum of the squared errors for the line, $y = 9x + 114$.

County	Index of Exposure	Cancer Deaths per 100,000	Best-fit y-values	Errors	Errors squared
Umatilla	2.49	147.1			
Morrow	2.57	130.1			
Gilliam	3.41	129.9			
Sherman	1.25	113.5			
Wasco	1.62	137.5			
Hood River	3.83	162.3			
Portland	11.64	207.5			
Columbia	6.41	177.9			
Clatsop	8.34	210.3			

Summary

Bivariate data from the real world is typically not perfectly linear like the **functions** studied in many math classes. This is why we say the **best fit line** *models* the bivariate data; it helps us see any trends and thus we use the term *trendline*. Occasionally we do encounter perfectly linear bivariate data, and this of course makes us happy.

As you read, jot down notes and questions. At the end of this section's Guided Worksheets there is a space for **Reflection** *and* **Monitoring Your Understanding***, where you can try and answer your questions after completing the guided activities and homework.*

Notes & Questions

Guided Practice Activity #1 – Tolls

The table shows the distances from Exit 24 (Albany) on the New York State Thruway and other exits, heading west towards Syracuse and Rochester (Exit 46).

Exit #	Miles	Toll	Exit #	Miles	Toll
24	0	0	34	114	$4.45
25	6	$0.25	35	131	$5.10
26	14	$0.60	38	138	$5.40
27	26	$1.00	40	157	$6.10
28	34	$1.35	41	173	$6.70
29	46	$1.80	42	180	$7.00
30	72	$2.80	43	193	$7.50
31	85	$3.30	44	200	$7.75
32	96	$3.75	45	204	$7.90
33	105	$4.10	46	215	$8.35

1. Is the toll proportional to the distance traveled? How do you know?

2. The scatterplot below does show a perfect linear fit between toll and distance traveled. It just does not go through the origin. Interpret the slope in real world terms.

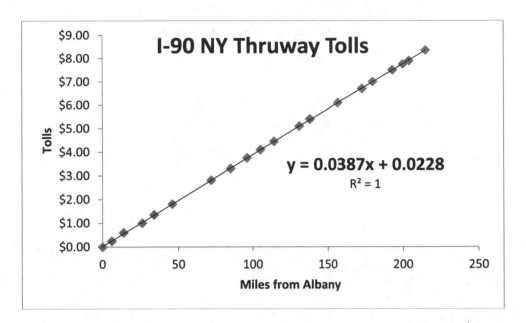

3. Interpret the y-intercept in real world terms.

Excel Preview – 7.1 Spurious

The Spurious Correlation website gives the following example of randomly falling data between the divorce rate in Maine and the per capita consumption of margarine in the U.S.

	Divorce Rate in Maine (divorces per 1000 people)	Per Capita Consumption of Margarine (pounds)
2000	5	8.2
2001	4.7	7
2002	4.6	6.5
2003	4.4	5.3
2004	4.3	5.2
2005	4.1	4
2006	4.2	4.6
2007	4.2	4.5
2008	4.2	4.2
2009	4.1	3.7

1. The following scatterplot shows an almost perfect linear fit! How do we know it is not perfectly linear?

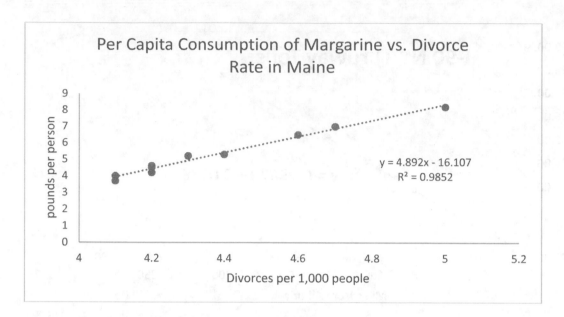

Per Capita Consumption of Margarine vs. Divorce Rate in Maine

y = 4.892x - 16.107
R² = 0.9852

pounds per person

Divorces per 1,000 people

2. Identify the points for the year 2000 and 2005.

3. Interpret the slope.

4. Has the per capita consumption of margarine been increasing or decreasing since 2000?

5. Why is this called a spurious correlation?

Section 8.2 Reflections
Monitor Your Understanding

Are you able to distinguish between a positive correlation and variables that decrease over time?

Section 8.3 Correlation Coefficient and Coefficient of Linear Determination

Objective 1 – Understand R and R^2
Objective 2 – Understand Best-fit Exponential Trendlines

Objective 1 – Understand R and R^2

Key Terms

Correlation coefficient, R	
Coefficient of linear determination, R^2	

Summary

We saw in the previous section a perfect linear fit between tolls and distance traveled. Note that the R^2 value was 1, which is telling us this is a perfect fit as we now discuss. In general, scatterplots of **bivariate data** will not be perfectly linear, as in all of the other scatterplots we have seen. Two persons looking at the same scatterplot can subjectively assess different "degrees of linearity" in the data. It would be helpful if we had a quantitative measure for a model's linearity, some kind of score we can attach to a set of data points. There is such a score. We are talking about a famous statistic called the **correlation coefficient**, or its very close cousin, the **coefficient of linear determination.** Since Excel will handle the calculations for us, we will skip directly to interpreting these measures and cover the formulas at the end of the chapter in the *Spotlight on Statistics*.

The correlation coefficient, denoted by the letter R, is a single number associated with bivariate data. It always falls between -1 and +1; that is, bivariate data can have an R-value of, say, 0.45 or -0.12, but never 3.14 or -50.

How R Measures the Strength of Relationship Between Two Variables				
Very Strong, Negative	Moderately Strong, Negative	Absolutely No Relationship	Moderately Strong, Positive	Very Strong, Positive
-1.0	-0.5	0	0.5	1.0

Notes & Questions

Guided Practice Activity #1 – R You Kidding Me!

1. The following scatterplot shows a time series plot of the price per barrel of oil. Sketch in a linear trendline and estimate the equation.

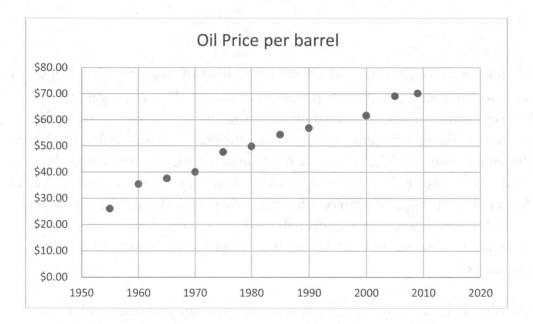

2. Interpret your slope using proper units and compare to the slope from the best fitting equation, $y = 0.7692x - 1474.1$.

3. Use this equation to predict oil prices in 2020.

4. Now consider the gas versus oil correlation. Interpret the slope of this trendline.

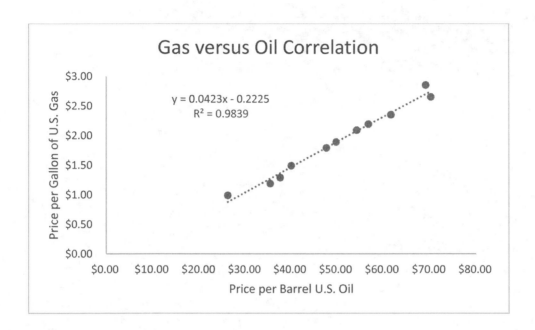

5. Interpret the R-squared value using: " $R^2 \cdot 100\%$ of the variability in the output can be attributed to variability in the input."

6. Use this equation to predict gas prices in 2020 using the answer to #3 above.

Now consider the relationship between strength of schedule (SOS) and the percentage of games won for women's Division III basketball.

7. Do you think this will be a positive relationship (harder the schedule more games won) or negative relationship? Explain!

8. What percentage of the variability in games won do you think is due to variability in the strength of schedule?

9. Check your guesses against actual data from the 2009-10 Division season:

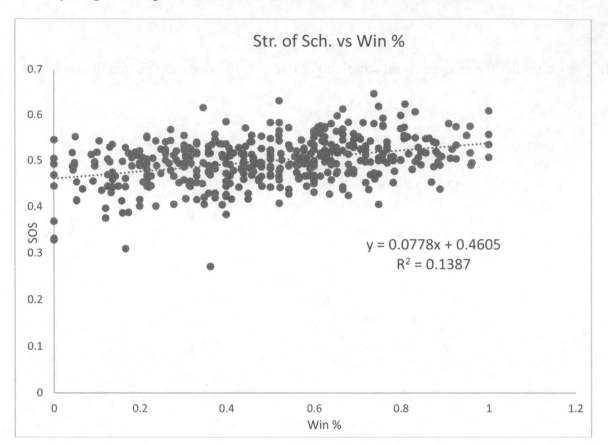

Str. of Sch. vs Win %

$y = 0.0778x + 0.4605$
$R^2 = 0.1387$

Summary

At this point you might be wondering how Excel can assign an R^2 value to an **exponential trendline** as we did in Chapter 6, since R^2 is a measure of linearity. The answer lies in the brief comment in the previous chapter on **logarithms** regarding **log scales**. When a data set is exponential, the logarithm of the outputs will be linear with respect to the inputs, and it is this linear fit which is assigned an R^2 value.

As you read, jot down notes and questions. At the end of this section's Guided Worksheets there is a space for **Reflection** *and* **Monitoring Your Understanding**, *where you can try and answer your questions after completing the guided activities and homework.*

Notes & Questions

Guided Practice Activity #1 – Learning Curve

A psychology student conducted a memory experiment using other students as subjects. The objective was to see how well the subject could memorize a random string of letters and how the length of the string affected the outcome. A subject was shown a string such as YGOPFGZ for five seconds, then had to wait 10 seconds, and then was asked to repeat the string. The data below summarizes the student's results. Given that the **percentage** of subjects correctly recalling the entire string can never go below zero, the student decided to fit her data with an exponential trendline.

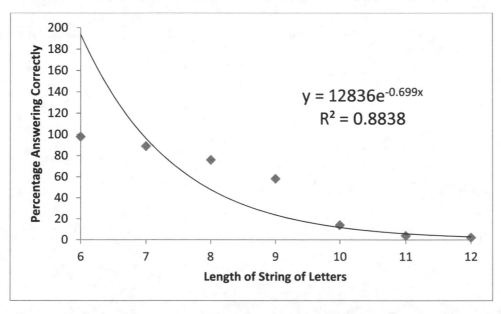

$$y = 12836e^{-0.699x}$$
$$R^2 = 0.8838$$

1. Interpret the continuous growth rate.

2. Compute the associated growth factor, $1 + r$.

3. Interpret the rate associated with the factor.

Excel Preview – 7.4 Wins

This is an example in which a relatively small correlation still carries significance. The reason is that the output variable is affected by a large number of factors, so when just one of them has a 45% coefficient of determination it gets our attention. Football coaches, fans and TV analysts love to try to reduce the winning of games to one or two simple variables: "Run and stop the run", "win the turn-over battle", "win the time-of-possession battle".

We explore the turn-over theory. A turn-over is a fumble or interception which changes which team possesses the ball, if you take the ball away from the other team you get a +1, but if you give the ball away to the other team you get a -1. Our input variable is the turn-over margin (positive turn-overs plus negative ones) for each NFL team during the entire 2017 regular season. The output variable is the number of victories (out of 16 games) for that team.

1. Interpret the slope of the best-fit line from the scatterplot below.

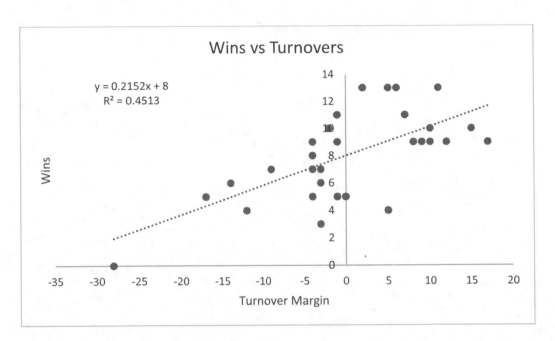

Wins vs Turnovers

y = 0.2152x + 8
R² = 0.4513

Wins

Turnover Margin

2. How many wins could a team expect with a turnover margin of 10? (3 decimal places)

3. Interpret the R-squared value.

4. Consider the data below from 2004. Have turnovers become more or less predictive of wins from 2004 to 2017?

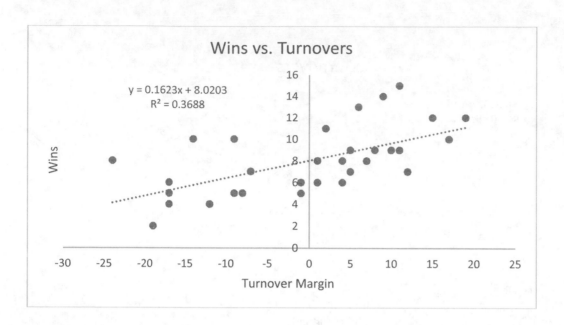

Wins vs. Turnovers

$y = 0.1623x + 8.0203$
$R^2 = 0.3688$

Section 8.3 Reflections
Monitor Your Understanding

Can you explain why R^2 is preferred over R?

Section 8.4 Spotlight on Statistics

Objective 1 – Understand Covariance and Correlation
Objective 2 – Explore $y = \overline{Y} + R \cdot \dfrac{SD_Y}{SD_X} \cdot (x - \overline{X})$

Objective 1 – Understand Covariance and Correlation

Key Terms

Variance	
Covariance	

Summary

In this last section we will formally define the **correlation coefficient** in terms of a related statistic, the **covariance**. Then we will derive the equation of best fit using intuitive notions of correlation and z-**scores**. To start, recall the formula for the **standard deviation**:

$$s = \sqrt{\frac{\sum (x_i - \overline{X})^2}{N-1}}$$

Squaring the standard deviation gives the variance of a data sample:

$$\text{Var} = \frac{\sum (x_i - \overline{X})^2}{n-1} = s^2$$

Both of these formulas are for **univariate data**; for **bivariate data** we compute the covariance:

$$\text{Covar} = \frac{\sum (x_i - \overline{X}) \cdot (y_i - \overline{Y})}{n-1}$$

CAUTION! Note these formulas are all assuming your data is a sample from a larger population. Replace $n-1$ with N if your data represents the entire population.

Note that the quotient of the **variance** and the **standard deviation** squared for **univariate data** is one:

$$\frac{\text{Var}}{s^2} = 1$$

This motivates (but does not prove!) the following inequality for **bivariate data**:

$$-1 \leq \frac{\text{Covar}}{s_x \cdot s_y} \leq 1$$

We have created a new statistic that is trapped between -1 and 1. Any guesses what this is? It's *R*! Excel has built-in functions for these statistics: **CORREL** will give the correlation coefficient, and **COVARIANCE.S** will give the sample covariance defined above.

As you read, jot down notes and questions. At the end of this section's Guided Worksheets there is a space for **Reflection** *and* **Monitoring Your Understanding***, where you can try and answer your questions after completing the guided activities and homework.*

Notes & Questions

Guided Practice Activity #1 – Oil and Gas Reprise

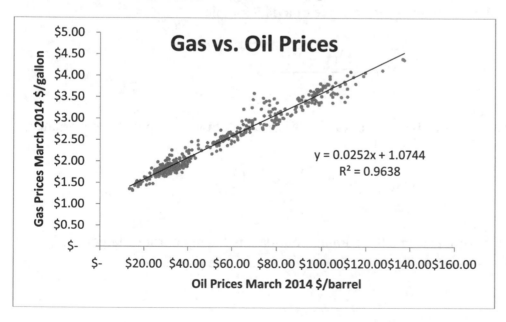

The scatterplot comes from the data in the following spreadsheet:

F3			×	✓	f_x	=CORREL(B3:B460,C3:C460)		

◢	A	B	C	D	E	F
1		Bivariate Data				
2	Month	Oil ($/barrel)	Gas ($/gallon)			
3	January 1976	$ 55.99	$ 2.55		CORREL	0.981718
4	February 1976	$ 55.85	$ 2.53		R^2	0.96377
5	March 1976	$ 56.80	$ 2.50			
6	April 1976	$ 56.20	$ 2.48		SD.oil	27.72543
7	May 1976	$ 55.98	$ 2.50		SD.gas	0.71207
8	June 1976	$ 55.98	$ 2.56		COVAR	19.38151
9	July 1976	$ 55.81	$ 2.57		R	0.981718
10	August 1976	$ 55.80	$ 2.58			
11	September 1976	$ 55.06	$ 2.58			

1. Recall the slope function: =**SLOPE**(known_*y*'s, known_*x*'s). If we switch the arguments does the slope change?

2. The correlation function has similar arguments but uses different words:
   =**CORREL**(array1, array2).
   If we switch the arguments does it change the correlation coefficient?

3. What formula is in cell **F4** to compute the coefficient of linear determination?

4. What is the standard deviation of the *x*-values?

5. What formula is in cell **F9** that computes the correlation coefficient using the covariance and standard deviations?

Summary

Well, the equation in this objective certainly looks daunting, but it will actually tie together much of the statistics we have been learning, going back to Chapters 1 and 2 on **standard deviation** and **z-scores**, and incorporating these concepts with the line of best fit from Chapter 5. The equation above is a general formula for the line of best fit. The basic idea is simple enough: we want to use the **line of best fit** from a scatterplot for **bivariate data** to make predictions.

We know from the previous guided activity that oil and gas prices are highly **correlated,** in fact $R^2 = 0.9638$ tells us that 96.38% of the variation in gas prices is due to variation in oil prices. Only 4% of the variability in gas prices can be attributed to such vagaries as the weather and the local gas station manager.

Thus knowing the oil price tells us a lot about the associated gas price. Consider $x = \$200$ for the oil price. It is clearly well above the average price for a barrel of oil in March 2014 \$, so we would expect the associated gas price to be well above its own **mean**. How far above the mean? Recall from Chapter 2, that we put a data value into context by talking about its **z-score**, or how many **standard deviations** above or below the mean the value is. If the oil price is five standard deviations above the mean, we would expect the gas price to also be five standard deviations above its own mean. Well, maybe not quite 5 since oil and gas are not **perfectly correlated**, but maybe 96%·5 standard deviations?

First we must compute the mean and standard deviation for both oil and gas prices:

March 2014 $	Oil Price per Barrel	Gas Price per Gallon
Mean	$54.28	$2.44
Standard Deviation	27.725	0.712

Now we can compute the z-scores for each:

$$z_{oil} = \frac{200 - 54.28}{27.725} = 5.256$$

$$z_{gas} = \frac{6.11 - 2.44}{0.712} = 5.154$$

Note that $\dfrac{z_Y}{z_X}$ in this example is $\dfrac{5.154}{5.256} = 0.98,$ which is the **correlation coefficient!**

CAUTION! The z-score for the output variable in **bivariate data** will be a fraction of the z-score for the input variable. The more correlated the bivariate data, the closer that fraction will be to 1. That fraction is R. Thus, $z_y = R \cdot z_x$, not R^2 as suggested above. We can also say that the z-scores are proportional, and the constant of proportionality is R.

As you read, jot down notes and questions. At the end of this section's Guided Worksheets there is a space for **Reflection** *and* **Monitoring Your Understanding***, where you can try and answer your questions after completing the guided activities and homework.*

Notes & Questions

Guided Practice Activity #1 – Fishy Data

The Fishy Data set consists of 42 length/weight measurements of trout from the Spokane River.

	A	B	C	D	E
1	Rainbow trout from the Spokane River, WA (n = 42)				
2	Source: WA State Dept. of Ecology report				
3	length (mm)	weight (g)		length (mm)	weight (g)
4	457	855		270	209
5	405	715		359	476
6	455	975		347	432
7	460	895		259	202
8	335	472		247	184
9	365	540		280	248
10	390	660		265	223
11	368	581		309	392
12	385	609		338	460
13	360	557		334	406
14	346	433		332	383
15	438	840		324	353
16	392	623		337	363
17	324	387		343	390
18	360	479		318	340
19	413	754		305	303
20	276	235		335	410
21	387	538		317	335
22	345	438		351	506
23	395	584		368	605
24	326	353		502	1300

Let X = length of trout (mm) and Y = weight of trout (g). The following statistics have been computed for you:

	Mean $\bar{X}_{length}$ or $\bar{Y}_{weight}$	Standard Deviation SD_X or SD_Y
X = length (mm)	352.90	57.02
Y = weight (g)	501.02	230.51

R_{XY}, the correlation coefficient, is 0.969.

1. Compute the least squares line using the equation:

$$y = \overline{Y} + R \cdot \frac{SD_Y}{SD_X} \cdot (x - \overline{X})$$

2. Interpret the slope as a rate of change.

3. Interpret the *y*-intercept.

4. Why does the y-intercept make no sense? Look at the scatterplot below. For what domain does this linear model seem valid?

5. You catch a whopper, measuring 484 mm in length! What is the z-score of this trout?

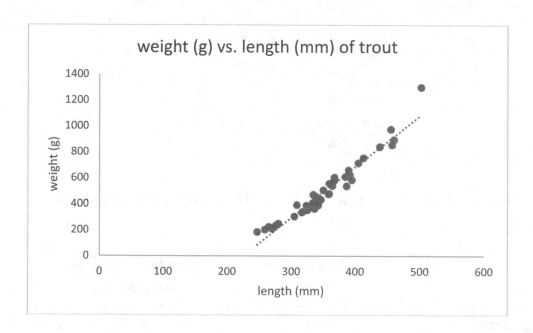

6. On the scatterplot mark the means, $\overline{X} = 352.9$ and $\overline{Y} = 501.02$, and the 484 mm on the respective axes.

7. What would you estimate the weight of this BIG fish to be? Place a mark on the y-axis for this weight.

8. What is the z-score of this weight? How is it related to the z-score of the length?

9. It turns out the z-scores are related by the following formula:
$$z_Y = R_{XY} \cdot z_X$$

Use this formula to compute the z-score of the weight and then compute the actual weight of the fish associated to the 484 mm.

Excel Preview – 8.4 Wins

We have looked at this Excel problem above, and now we will compute the equation of the best-fit line using the techniques of this section.

	A	B	C	D
1			2017	
2			Turnovers	Wins
3		Baltimore	17	9
4		Kansas City	15	10
5		LA Chargers	12	9
6		Philadelphia	11	13
7		Detroit	10	9
8		Jacksonville	10	10
9		Buffalo	9	9

We can compute the statistics required using our built-in Excel functions:

	CORREL	0.672
	R²	0.451

	Turnovers	Wins
Mean	0.000	8.000
StdDev	10.000	3.203

1. Compute the least squares line using the equation and having wins be the y-values:
$$y = \overline{Y} + R \cdot \frac{SD_Y}{SD_X} \cdot (x - \overline{X})$$

2. Compute the z-score for $X = 10$ turnover margin.

3. Compute the z-score for the number of wins for a team with a turnover margin of 10 using:
$$z_y = R \cdot z_x$$

4. Compute the expected number of wins for a team with 10 turnovers using the formula for z-score.

5. Compute the expected number of wins for a team with 10 turnovers using the best-fit equation.

Section 8.4 Reflections
Monitor Your Understanding

Can you explain the connection between z-scores and the best-fit line?

Chapter 9 – Financial Health

Section 9.1 Retirement

Objective 1 – Plan for Retirement

Key Terms

Securities	
Social security	
Medicare	

Summary

We begin our chapter on financial health by looking at how to save for **retirement**. Investments are called **securities** since they make you feel secure knowing you have money available for retirement or other life choices. Unfortunately not everyone is saving money, as we saw in **Figure 6.10** on the Personal Savings Rate and as also shown in **Table 9.1**.

TABLE 9.1 Before-Tax Family Income and Percentage of Families That Saved by Income Percentile 2013 & 2016

	2013				2016			
	Income		Percentage of families that saved	Percentage of families	Income		Percentage of families that saved	Percentage of families
Family characteristic	Median	Mean			Median	Mean		
All families	48.1	89.9	53.0	100.0	52.7	102.7	55.4	100.0
Percentile of income								
Less than 20	14.2	13.7	31.7	20.0	15.1	14.4	32.1	20.0
20–39.9	29.3	29.4	40.9	20.0	31.4	31.8	45.2	20.0
40–59.9	48.1	48.7	49.6	20.0	52.7	53.4	57.2	20.0
60–79.9	78.8	81.0	64.5	20.0	86.1	87.4	64.8	20.0
80–89.9	125.5	128.2	73.3	10.0	136.0	138.7	72.8	10.0
90–100	236.8	424.8	83.4	10.0	260.2	514.7	82.3	10.0

Note: In thousands of 2016 dollars except as noted
Data from: 2016 Survey of Consumer Finances, Historic Tables and Charts, Table 1, https://www.federalreserve.gov/econres/scfindex.htm

Note that only 55.4% of families saved in 2016, but that was an increase from 53% in 2013, along with an increase in average household income (median and mean). A lower percentage (32.1%) of families in the lowest quintile of income was able to save, as would be expected. It is difficult to save if you are making less than a living wage, which is why raising the minimum wage is so critically important to many families. You may be interested in knowing the different types of financial assets people are investing their savings in:

TABLE 9.3
Value of Financial Assets of All Families Distributed by Percent, 2001–2016

Transaction accounts are checking and savings accounts. Note the large increase in "pooled investment funds" or "mutual funds" in 2016.

Type of financial asset	Value of Financial Assets Distributed by Percent					
	2001	2004	2007	2010	2013	2016
Transaction accounts	11.4	13.1	10.9	13.3	13.3	11.8
Certificates of deposit	3.1	3.7	4.0	3.9	2.0	1.5
Savings bonds	0.7	0.5	0.4	0.3	0.3	0.2
Bonds	4.5	5.3	4.1	4.4	3.2	2.8
Stocks	21.5	17.5	17.8	14.0	15.9	13.7
Pooled investment funds (excluding money market funds)	12.1	14.6	15.8	15.0	14.8	23.2
Retirement accounts	29.0	32.4	35.1	38.1	38.8	35.6
Cash value life insurance	5.3	2.9	3.2	2.5	2.7	2.2
Other managed assets	10.5	7.9	6.5	6.2	7.6	7.7
Other	1.9	2.1	2.1	2.3	1.5	1.3
Total	100	100	100	100	100	100.0
Financial assets as a share of total assets	42.2	35.8	34.0	37.9	40.8	42.5

Data from: 2016 Survey of Consumer Finances, Historic Tables and Charts, Table 5, https://www.federalreserve.gov/econres/scfindex.htm

A note of caution: three-fourths of Americans age 55 to 64 years old have less than $30,000 saved for retirement. These pre-retirees are at risk of not being able to maintain their standard of living in retirement and have to work longer. Retiring with enough money to live on comfortably is within reach of all Americans, it simply requires a disciplined savings and investing strategy starting from the day you begin work. Then you just sit back and let the power of **compound interest** and **exponential growth** take over!

As you read, jot down notes and questions. At the end of this section's Guided Worksheets there is a space for **Reflection** *and* **Monitoring Your Understanding***, where you can try and answer your questions after completing the guided activities and homework.*

Notes & Questions

Guided Practice/Excel Activity #1 – How Retirement Works

We will spend much of the chapter going over how you save the $1,000,000 shown below for retirement. It is important to know how you will use this million dollars when you retire, and this activity walks you through the basics.

	A	B	C	D	E	F	G	H
1	Initial Savings	$ 1,000,000		Year	Balance	Withdrawal	New Balance	End Balance
2	Withdrawal Rate	5%		1	$1,000,000	$ 50,000	$ 950,000	$ 1,007,000
3	Inflation Rate	3%		2	$1,007,000	$ 51,500	$ 955,500	$ 1,012,830
4	Rate of Return	6%		3	$1,012,830	$ 53,045	$ 959,785	$ 1,017,372
5				4	$1,017,372	$ 54,636	$ 962,736	$ 1,020,500
6				5	$1,020,500	$ 56,275	$ 964,224	$ 1,022,078
7				6	$1,022,078	$ 57,964	$ 964,114	$ 1,021,961
8				7	$1,021,961	$ 59,703	$ 962,258	$ 1,019,994

This Excel spreadsheet maps out your annual withdrawals for 25 years from a **retirement** account starting with the given initial savings and a 5% withdrawal rate that increases with **inflation**. Assume an inflation rate of 3% and a **rate of return** on your investment of 6%, and determine how long your money will last.

1. The *Withdrawal* at first is a percentage of the year's starting balance using the withdrawal rate. Then it grows each year using the inflation rate. What formula is in cell **F2**?

    a.   What formula is in cell **F3** that can be filled down?

2. The *New Balance* is what is left in your account after subtracting the withdrawal from the balance. What formula is in cell **G2** that can be filled down?

3. The *End Balance* is now the value of your account after it has grown for 1 year at the rate of return. What formula is in cell **H2** that can be filled down?

4. The *Balance* should link to the Initial Savings at first, then it will link to the end balance of the previous year. What formula is in cell **E3** that can be filled down?

5. If the rate of return doubles from 6% to 12%. Does the amount after 25 years also double?

Section 9.1 Reflections
Monitor Your Understanding

What did you learn in this section that was surprising?

Section 9.2 Securities: Cash, Stocks, Mutual Funds, and Bonds

Objective 1 – Understand Certificates of Deposit
Objective 2 – Understand Bonds
Objective 3 – Understand Stocks
Objective 4 – Understand Mutual Funds

Objective 1 – Understand Certificates of Deposit

Key Terms

Certificates of deposit	
Checking/Savings Accounts/Money Market Accounts	
Discount rate	
Federal funds rate	
Prime rate	

Summary

Now that we have a better sense of why saving for **retirement** is important and how you can use your retirement account to generate income, let's figure out how to get the $1,000,000 we saw in the previous section into your retirement account. Looking back at **Figure 9.3**, the first two financial assets mentioned are "Transaction Accounts" and "**Certificates of Deposits**". Transaction accounts refer to checking and savings accounts and are the first step to financial security. Unfortunately 25% of all U.S. households were either underbanked or unbanked according to the 2010 U.S. Census.

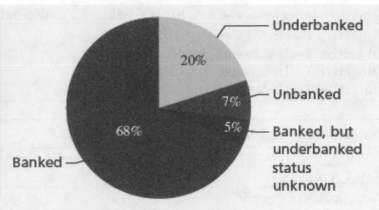

Unbanked means the household does not have a checking or savings account. Underbanked refers to households that do have such accounts but also rely on money orders, pawn shops, check-cashing services, rent-to-own agreements, or payday loans. All these services should be avoided literally at all costs. They are a financial plague to the poor and financially illiterate.

FIGURE 9.4 Banking Status of U.S. Households, 2015
Note: Percentages are based on 127.6 million U.S. households; based on data from FDIC.
Data from: Federal Deposit Insurance Corporation 2015 Survey https://www.fdic.gov/householdsurvey/2015/2015execsumm.pdf

Checking and savings accounts are basically a cash investment. It is always nice to keep cash on hand, and there are several ways to keep your money safe and earn a modest return. The Federal Deposit Insurance Corporation (FDIC) guarantees the money that you deposit in a bank account up to a specific amount. FDIC insurance covers all deposit accounts, including checking and savings accounts, money market deposit accounts, and certificates of deposit. The standard insurance amount is $250,000 per depositor, per insured bank, for each account ownership category (as of 2018). If you have more cash than this you can simply open accounts at different banks.

Checking/savings/money market accounts are ultra-safe investments and correspondingly offer low **rates of return**. There are slight differences between savings and money market accounts, most noticeably money markets often have checks. In 2018 the average **APY** was typically below 0.5% at most banks with the best offering 1.75% with no minimum deposit. If **inflation** is 3% a year these accounts have a negative real rate of return. You can withdraw your money at any point with no penalty.

Certificates of Deposits (CDs) require you to invest your cash for a fixed period of time and give a fixed APY. Withdrawing early will result in penalties. The longer you guarantee the bank your cash, the better the return. Again, ultra-safe with corresponding negative real rate of return.

As you read, jot down notes and questions. At the end of this section's Guided Worksheets there is a space for **Reflection** *and* **Monitoring Your Understanding**, *where you can try and answer your questions after completing the guided activities and homework.*

Notes & Questions

<u>Guided Practice Activity #1</u> – APR vs APY in a CD

You buy a $50,000 jumbo CD (Certificate of Deposit) with an APR of 2% compounded bi-annually. This is called a *6-month CD*, you are allowed to withdraw your funds at the end of any 6-month period, but not before without a penalty (they promise to take good care of your money and water it every day).

1. How many periods are there in 1 year?

2. What is the periodic rate?

3. How much interest do you gain in the first period?

4. How much money will you have at the end of 1 year?

5. What is the APY?

Key Terms

Savings bonds	
Bonds	
Coupon rate	
Treasuries	
Treasury bills (T-bills)	
Treasury notes (T-notes)	
Treasury bonds (T-bonds)	
Municipal bonds (Munis)	
Corporate bonds	

Summary

Savings bonds have a similar **rate of return** as savings accounts and **CDs**. The difference is that now you are in effect lending money to the Federal Government to pay its bills. In return they will give you interest on the amount you loan them (i.e. the value of the bond). There are two types of savings bonds, I bonds which have an **interest rate** that varies with **inflation**

(2.52% composite rate in July 2018) and EE bonds which come with a fixed rate (0.10% in April 2018). In addition to **Savings Bonds**, the Federal Government "sells" **Treasuries** which come in three flavors, depending on the **term**. You can buy them online through treasurydirect.gov, or through a bank or broker. They are sold in denominations of $100.

Bonds are essentially loans to either the Federal Government (**Treasuries**), local governments or municipalities (**Munis**), or to companies (**Corporate Bonds**). All three of these types of bonds traditionally pay simple interest semiannually. Bonds are rated by independent agencies based on the bond issuers' ability to pay back the loan, in just the same way people are given credit scores. Bonds are rated from best (AAA) to worst (C or junk bonds). Bond holders thus have a lender/creditor stake.

Every investment has its risks; the more risk, the higher the rate of return. **Treasuries** are considered the safest investment in the world since they are backed by the U.S. government. **Corporate bonds** carry the risk of default by the company issuing the **bond**. For example, on June 1, 2009, General Motors filed for bankruptcy for a variety of reasons, foremost among them their inability to fund their pension obligations to all their retirees. At the time of bankruptcy there was roughly $29 billion in GM bonds being held. The bankruptcy settlement, completed in April of 2011, allotted bondholders about 10% of the new GM (worth about $4 billion) by giving them **shares of stock** in the new company. This roughly translated into bondholders receiving back only 30 cents for every dollar they loaned GM (i.e. a 70% loss). Bonds can also lose value if **interest rates** rise.

As you read, jot down notes and questions. At the end of this section's Guided Worksheets there is a space for **Reflection** *and* **Monitoring Your Understanding**, *where you can try and answer your questions after completing the guided activities and homework.*

Notes & Questions

Guided Practice Activity #1 – Fixed Income

The company you work for is selling $10,000 bonds at 8% for 5 years. This means you will get *simple interest* every 6 months for the 5 year term. You buy 1 bond.

1. What is the face value?

2. What is the coupon rate?

3. What is the periodic rate?

4. How much will the company pay you every 6 months?

5. Why are bonds referred to as *fixed income* investments?

6. How much will the company pay you over the entire term (including final payment of face value)?

7. Use the geometric mean to determine the *average annual rate of return* for this investment over the 5 years.

Guided Practice Activity #2 – Bond Risk

You buy a $10,000 bond at 4% for 15 years but need to sell the bond after only 6 years due to unexpected events in your life.

1. How much would someone pay for this bond if interest rates have risen to 6%?

2. Use the geometric mean to determine the average annual rate of return for this investment over the 9 years they own the bond.

3. How much would someone pay for this bond if interest rates had dropped to 3% instead of rising to 6%?

4. Looking at the chart of the Federal Funds Rate below, which direction were the rates heading in 2018?

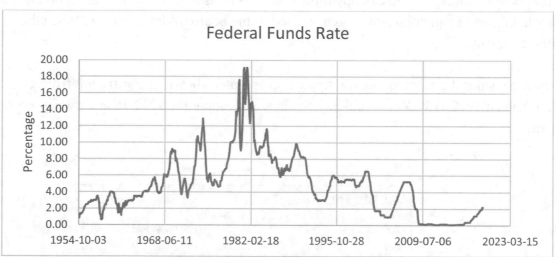

Data from: Board of Governors of the federal Reserve System https://fred.stlouisfed.org/series/FEDFUNDS

Key Terms

S&P 500	
Ticker symbol	
Market capitalization	

Summary

So far we have covered the safest investments, but these have not had very high **rates of return**. **Stocks** are shares of a company which you can buy through an online brokerage account like E*trade and TD Ameritrade. When you buy shares of a company, you have an ownership/equity stake in the company. This means you get to vote at shareholder meetings, with number of votes equal to number of shares. If you can buy more than half the stock in the company you have a controlling interest and can control who gets elected to the board of directors, and thus who controls the company.

For example, consider the stock quote for Apple Incorporated shown in **Figure 9.7** from Google Finance on August 14, 2018. You could buy one share of Apple for $209.75 and own a piece of the company.

> Note that every company with publicly traded stock has a **ticker symbol**. The symbol for General Motors is GM, and for Apple it is AAPL.

FIGURE 9.7 Apple Inc. Stock Quote
Data from: Google Finance, www.google.com

CAUTION! Buying stock in just one company can be very risky. It is much easier to buy a mutual fund for the entire S&P 500 which essentially gives you ownership of 500 companies and diversifies your risk.

As you read, jot down notes and questions. At the end of this section's Guided Worksheets there is a space for **Reflection** *and* **Monitoring Your Understanding***, where you can try and answer your questions after completing the guided activities and homework.*

Notes & Questions

Guided Practice Activity #1 – Start-up

A friend is starting a website company and is selling 10,000 shares at $15 each; you buy 1,000 shares. At the end of the year the company makes a profit of $12,000, and decides to reinvest half into the business and payout the rest as *dividends* to the owners. Determine the following statistics:

1. Market capitalization:

2. Share price:

3. Earnings per share (EPS):

4. Price to Earnings (PE ratio):

5. Dividends (per share):

6. Dividends (as % of share price):

7. Someone offers to buy your shares from you for $20 a share. Do you sell?

Objective 4 – Understand Mutual Funds

Key Terms

Mutual fund	
Net asset value (NAV)	
Index fund	
Expense fee	
Front end load	

Summary

Choosing your own **stocks** to buy can be intimidating and it can also be difficult to *diversify* your holdings. If you have only $1,000 to invest, you most likely will buy shares of only one or two companies. Remember, an online brokerage account charges you $9.95 per order (approximately), so if you buy twenty shares of one company at $50 you get charged $9.95, but if you buy one share of twenty companies (twenty separate orders) you get charged twenty times $9.95 = $199. A **mutual fund** is a pool of money from many investors that is invested for them by a manager who charges a fee for this service.

One of the greatest investors of our time, Warren Buffett, CEO of Berkshire Hathaway, the "Oracle of Omaha," and third wealthiest individuals in the world (2018) with a net worth of $84 billion, offers us very simple investing advice, telling us we don't need to be experts to achieve satisfactory returns:

> "If 'investors' frenetically bought and sold farmland to one another, neither the yields nor the prices of their crops would be increased. So ignore the chatter, keep your costs minimal, and invest in stocks as you would in a farm. Put 10% of cash in short-term government **bonds** and 90% in a very low-cost **S&P 500** index fund."

Then put your faith in the American economy and let your money grow. Constantly digging up your seeds and replanting them is a good way to destroy your crop. Read Warren Buffet's article, *Why I Like to Think of Stocks Like Farms*, **Fortune** Magazine March 17, 2014. Then read the rest of the magazine and continue to read it every month. **Money** Magazine is another good financial read every month. If you seriously want to grow your money, then you need to get savvy and do something about it.

As you read, jot down notes and questions. At the end of this section's Guided Worksheets there is a space for **Reflection** *and* **Monitoring Your Understanding***, where you can try and answer your questions after completing the guided activities and homework.*

Notes & Questions

Guided Practice Activity #1 – Mutually Funding

You invest $25,000 in a mutual fund with a 1% load and 2.3% expense fee. The fund increases by 18.6% over the course of the year.

1. What is the front end load they charge you?

2. How much money is invested in the fund?

3. How much does your investment grow by in 1 year?

4. How much is the expense fee? (They charge 2.3% of the account value.)

5. How much is your investment worth at the end of one year?

You are going to put money away every year into a mutual fund which returns 10% a year. You will deposit $6,000 in the first year and then increase your deposit by 3% every year (since your salary will be increasing also). Assume you start with $0 in the account and make the deposit at the beginning of the year so it always grows by 10% each year.

	A	B	C	D	E
1	Rate of Return:	10%			
2			a.)	a.)	a.)
3		Year	New Balance	Deposit	End Balance
4		1	$ -	$ 6,000.00	$ 6,600.00
5		2	$ 6,600.00	$ 6,180.00	$ 14,058.00
6		3	$ 14,058.00	$ 6,365.40	$ 22,465.74
7		4	$ 22,465.74	$ 6,556.36	$ 31,924.31
8		5	$ 31,924.31	$ 6,753.05	$ 42,545.10

1. What formula is entered in cell **C5** that can be filled down?

2. What formula is entered in cell **D5** that can be filled down?

3. What formula is entered in cell **E5** that can be filled down?

4. Assume it takes 25 years for this investment to be worth $1 million. Determine what the value of that $1 million is in today's dollars by assuming a 3% inflation rate. In other words, what amount of money today would grow at 3% for 25 years to equal 1 million?

Section 9.2 Reflections
Monitor Your Understanding

Can you explain why people invest in mutual funds versus buying stocks on their own?

Section 9.3 401(k) Plans

Objective 1 – Explore Saving for Retirement

Key Terms

401(k)	

Summary

To save for **retirement,** you need to open a retirement account, which generally is a **401(k),** named after the legislation which created this type of account. *Money* Magazine's article, *Building the New Retirement*, in the March 2014 issue gives us some history of the 401(k). In 1978 almost half of workers had pensions, a fixed salary to be paid to them during retirement from their company (or government). A new pension law passed in this year, subsection 401(k) of the Internal Revenue Code, made 401(k)s possible. Basically employers can offer a retirement plan with a defined annual contribution to your retirement account which is deducted from your paycheck before taxes (up to $18,500 in 2018). Employers can also offer to match a fraction of your contribution. Typically employers will offer to match 50 cents for every dollar you contribute up to 6% of your salary.

CAUTION! This means the company will contribute a maximum of 3% (not 6%) of your salary each year. If you make $100,000 and only contribute $4,000 then they match $2,000; if you contribute $8,000 they match $3,000 (the max). Unfortunately many workers do not take advantage of this free money. You should always max out your employer match.

As you read, jot down notes and questions. At the end of this section's Guided Worksheets there is a space for **Reflection** *and* **Monitoring Your Understanding***, where you can try and answer your questions after completing the guided activities and homework.*

Notes & Questions

Guided Practice Activity #1 – Comparing 401(k)'s

Three friends, Sal, Cal, and Val, all work for the same company and all earn the same salary of $40,000. The company will match half of their contributions to the **401(k) plan** up to 6% of their salary. Sal decides to contribute $100 a month to his 401(k), Cal decides not to contribute anything and invest 6% of what is left of his salary after taxes on his own, and Val decides to contribute $400 a month to her 401(k).

1. Determine what 6% of their salary is on a monthly basis.

2. What is the maximum the company will contribute to their 401(k)'s on a monthly basis?

3. How much does the company contribute to Sal's account on a monthly basis?

4. How much in total does Sal invest over the year?

5. How much does the company contribute to Val's account on a monthly basis?

6. How much in total does Val invest over the year?

7. Assuming Cal is taxed at a rate of 25%, how much of his salary is left after taxes?

8. How much does Cal invest in one year?

9. Compare how much Cal and Sal both spent on their annual investments.

Section 9.3 Reflections
Monitor Your Understanding

Can you explain what a company match means for a 401(k)?

Section 9.4 Built-in Financial Functions

Objective 1 – Use Rate, NPER, PMT, PV and FV (RNPPF)

Key Terms

RATE/NPER/PMT/PV/FV	

Summary

In the beginning of this chapter, we explored how to model withdrawing from your **retirement account** using a spreadsheet. Excel has built-in financial **functions** that can greatly speed up these "what-if" analyses.

CAUTION! You must enter any value that comes out of your pocket (like a payment to the bank or your retirement account) as a negative. Any value like a loan that you borrow from the bank is positive, since this is money going into your pocket.

CAUTION! RATE gives you the periodic rate not the APR; you must multiply by the number of periods in one year to get the **APR**. NPER gives the number of periods for the entire term of the investment/loan, not the number of **periods** in one year.

As you read, jot down notes and questions. At the end of this section's Guided Worksheets there is a space for **Reflection** *and* **Monitoring Your Understanding**,*where you can try and answer your questions after completing the guided activities and homework.*

Notes & Questions

Guided Practice Activity #1 – Rover Needs Poopy Paper Fast (RNPPF)

Use the **RATE, NPER, PMT, PV** and **FV** functions to answer the following. Remember the following important subtleties:

- RATE = periodic rate; you must divide the APR by the number of periods in one year.
- Anytime you send money to the bank it must be entered as a negative.
- It is assumed all PMT's are made at the END of each period, while PV is the amount you have to START. (You may use the TYPE argument to change PMT's to the beginning of the period.)

Write down the function as you would type into Excel:

1. What is your monthly payment on a $23,000 car loan at 8.75% for 5 years?

2. What APR will you need to guarantee yourself an annual payment of $60,000 for 25 years if you have $1,200,000 in the bank and you anticipate having nothing left at the end of 25 years?

3. How much money will you end up with if you make $650 monthly payments to your retirement account which earns 11.3% for 36 years?

4. Compute the following without typing them into Excel:

    a. = **FV** (5%, 2, 0, 10000)

    b. = **NPER** (8%, 0, -500, 583.2)

c. = **RATE** (2, 0, -1000, 1210)

d. = **FV** (4%, 3, -1000, -500)

	☐	☐	☐	☐
☐	☐	☐	☐	☐
☐		☐	☐	☐
☐			☐	☐
☐				☐

Excel Preview – 9.6 Payment

This problem will compute the monthly payments associated to a $20,000 car loan for various rates and terms.

◢	A	B	C	D	E
1	Loan:	$20,000			
2					
3		a.)	a.)	a.)	a.)
4	Years	Monthly Payments			
5		1%	4%	7%	10%
6	3	$564.16	$590.48	$617.54	$645.34
7	4	$425.23	$451.58	$478.92	$507.25
8	5	$341.87	$368.33	$396.02	$424.94

1. What formula can you enter in cell **B6** that can be filled down and across? Note we are making the payments positive outputs.

2. In this next problem we will use the screenshot from the examples in the text. You deposit
   $450 a month into a **mutual fund** in your **401(k)** which has averaged a 10% annual **rate of
   return**.

	A	B	C	D	E
1	APR:	10%	RATE:	0.83%	
2	Years:	30	NPER:	360	
3			PMT:	($450.00)	
4			PV: $	-	
5			FV:	$1,017,219.57	
6					
7	PMT:	($450.00)			
8				APR	
9	Years	6%	8%	10%	12%
10	10	$73,745.71	$82,325.72	$92,180.24	$103,517.41
11	20	$207,918.40	$265,059.19	$341,715.98	$445,164.91
12	30	$452,031.77	$670,661.75	$1,017,219.57	$1,572,733.86
13	40	$896,170.83	$1,570,953.52	$2,845,835.81	$5,294,147.63
14					

a. What formula is inc ell D1 to compute the periodic rate?

b. What formula can be entered into cell **D5** to compute the future value of this investment?

c. What formula can be entered into cell **B10** that can be filled down and across to compute
    the future value of these investment scenarios?

Section 9.4 Reflections
Monitor Your Understanding

Section 9.5 Loans

> **Objective 1** – Explore Credit Cards
> **Objective 2** – Understand Amortized Loans

> **Objective 1** – Explore Credit Cards

Summary

In this last section we will explore the flip side of investing, i.e. borrowing money or taking out a loan. Loans in and of themselves are not necessarily bad for you financially. It makes sense to take out a **mortgage** (loan to buy a house) at 4%, if your money is invested at 10%. Credit cards offer convenience and are a good way to build your credit score. The trick is to avoid fees and interest by paying off the credit card in a timely manner. Consider the following actual credit card offer.

TABLE 9.9 Credit Card Offer

Annual Percentage Rate (APR) for Purchases	0% introductory APR through your 06/2018 billing period. After that, your APR will be 10.9%, 14.9%, or 18.9% based on your creditworthiness. This APR will vary with the market based on the Prime Rate.
APR for Transfers	0% introductory APR through your 06/2018 billing period. After that, your APR will be 10.9%, 14.9%, or 18.9% based on your creditworthiness. This APR will vary with the market based on the Prime Rate.
APR for Cash Advances	24.9%. This APR will vary with the market based on the Prime Rate.
Penalty APR and When It Applies	29.4%. This APR will vary with the market based on the Prime Rate. This APR may be applied to your account if you make a late payment. How long will the penalty APR apply? If APRs are increased for a payment that is more than 60 days late, the Penalty APR will apply indefinitely unless you make the next six consecutive minimum payments on time following the rate increase.
Paying Interest	Your due date is at least 25 days after the close of each month. We will begin charging interest on cash advances on the transaction date.

CAUTION! Note the three possible **interest rates** you can be charged based on your credit score. It is crucial you maintain good credit by paying bills on time.

The credit card offer displays three possible **interest rates** depending on your credit score: 10.9%, 14.9%, and 18.9%. Credit cards are different from loans for a car, education, or a home because there is not a mandatory fixed payment due each month. Credit cards have a minimum payment due based on your balance, but you are free to pay whatever you wish above and beyond the minimum. In addition there is a not fixed term, or length of time, in which to pay back the loan. Car loans, student loans, and home loans have a fixed monthly payment and a fixed term in which to pay off the loan. Loans of this type are called amortized.

As you read, jot down notes and questions. At the end of this section's Guided Worksheets there is a space for **Reflection** *and* **Monitoring Your Understanding***, where you can try and answer your questions after completing the guided activities and homework.*

Notes & Questions

Guided Practice Activity #1 – Cancun

You charge a spring break trip to Cancun with your friends and come back with a nice tan, many happy memories, and a balance of $1,652 on your credit card. Compute how long it will take to pay off the card assuming you can make a $100 payment each month, for each of the three **APR**s: 7.9%, 14.9%, and 28.9%. Compute the total **interest** paid in each scenario.

1. We wish to compute how long to pay off our balance so which of the five functions (RNPPF) should we use?

2. How would you type this function into cell **B3** below that can be filled across?

◢	A	B	C	D
1			APR	
2		7.9%	14.9%	28.9%
3	NPER	17.55	18.60	21.32
4	Interest	$102.68	$208.33	$479.53

3. What formula is in cell B4 to compute the interest?

4. How much more in interest is paid with the worst rate compared to the best rate?

Objective 2 – Understand Amortized Loans

Key Terms

Mortgage	
Amortized loan	
Amortization schedule	
Mortgage rate	
Closing costs	

Summary

An **amortized loan** has a fixed monthly payment and fixed **term**. In Chapter 1 we looked at car loans and used the complicated payment formula. Now let's consider a home loan, called a **mortgage**, and use the **PMT** function to build a payment schedule, called an **amortization schedule**, detailing each month's payment and **interest**. Note the root "mort" in amortize, as in mortality or death. You are gradually killing off your debt.

CAUTION! The Monthly Payment will be negative because we used the PMT function, and anything coming out of your pocket is a negative.

When researching mortgage rates you see two different **interest rates**, the Mortgage Rate and the **APR**. The Mortgage Rate assumes the costs associated with getting a mortgage are included in the **Principal**, the APR does not.

CAUTION! Closing costs are associated with every mortgage and must be taken into account when choosing a mortgage.

*As you read, jot down notes and questions. At the end of this section's Guided Worksheets there is a space for **Reflection** and **Monitoring Your Understanding**, where you can try and answer your questions after completing the guided activities and homework.*

Notes & Questions

Guided Practice Activity #1 – Amortization Schedule

The following shows an amortization schedule for a 30 year mortgage on an initial loan of $200,000 at 6% APR:

	A	B	C	D	E	F
1	Principal:	$ 200,000				
2	APR:	6.00%				
3	Term:	30	years			
4						
5	Month	Beginning Balance	Monthly Payment	Interest Payment	Principal Payment	Ending Balance
6	1	$200,000.00	($1,199.10)	$ 1,000.00	($199.10)	$199,800.90
7	2	$199,800.90	($1,199.10)	$ 999.00	($200.10)	$199,600.80
8	3	$199,600.80	($1,199.10)	$ 998.00	($201.10)	$199,399.71
363	358	$ 3,561.63	($1,199.10)	$ 17.81	($1,181.29)	$ 2,380.33
364	359	$ 2,380.33	($1,199.10)	$ 11.90	($1,187.20)	$ 1,193.14
365	360	$ 1,193.14	($1,199.10)	$ 5.97	($1,193.14)	$ (0.00)

Write down the formulas in each of the following cells:

1. **B6** (Beginning Principal)

2. **C6** (Monthly Payment)

3. **D6** (Interest Payment)

4. **E6** (Principal Payment)

5. **F6** (Ending Principal)

6. **B7** (Beginning Principal)

Now fill down the formulas in row 6, **C6:F6**, into row 7. Then highlight all cells in row 7, **B7:F7**, and fill down!

7. How much interest do you pay over the 30 years? How does lowering the APR to 5% (new **PMT** = $1,073.64) change this answer?

Excel Preview – 9.5 Credit Card

Assume you have a $3,500 balance on a credit card with a 14.5% APR. You are only able to make $150 payments each month and you get an offer for a new credit card which will give you 0% on transferred balances for 6 months and then switch to a 16.7% rate. They will charge you 3% on any balance you transfer.

	A	B	C	D	E	F
1	APR:	14.50%				
2			a.)		a.)	a.)
3		Month	Balance	Payment	Interest	New Balance
4		1	$3,500.00	$ 150.00	$ 42.29	$ 3,392.29
5		2	$3,392.29	$ 150.00	$ 40.99	$ 3,283.28
6		3	$3,283.28	$ 150.00	$ 39.67	$ 3,172.95
28		25	$ 521.84	$ 150.00	$ 6.31	$ 378.15
29		26	$ 378.15	$ 150.00	$ 4.57	$ 232.72
30		27	$ 232.72	$ 150.00	$ 2.81	$ 85.53
31		28	$ 85.53	$ 150.00	$ 1.03	$ (63.44)
32		29	$ (63.44)	$ 150.00	$ (0.77)	$ (214.20)
33		30	$ (214.20)	$ 150.00	$ (2.59)	$ (366.79)
34						

1. What formula is in cell **D4** that can be filled down?

2. What formula is in cell **E4** that can be filled down?

3. What formula is in cell **F4** that can be filled down?

4. What formula is in cell **C5** that can be filled down?

5. Why doesn't the balance hit zero exactly in our spreadsheet?

6. How much in total does it cost to pay off the credit card?

7. If you switch to the new card what will be your initial balance?

Section 9.5 Reflections
Monitor Your Understanding

What is the most important thing you have learned about saving for retirement?

Chapter 10 – Logically!

Objective 1 – Understand Logic Vocabulary

Key Terms

Conditional statement	
Converse statement	
Contrapositive statement	
Inverse statement	
Negation	
Converse error	
Logically equivalent	

Summary

The foundation of logic is the *If-Then* statement, called a **conditional** statement. These statements in textbooks tend to be straightforward: If it is sunny then I will go to the beach; but in reality it is not always clear what the conditional statement is. The quotation in the text from *Alice in Wonderland* is filled with conditionals, can you find them?

> *"Then you should say what you mean," the March Hare went on. "I do," Alice hastily replied; "at least—I mean what I say—that's the same thing you know."*
> *"Not the same thing a bit!" said the Hatter. "Why, you might just as well say that 'I see what I eat' is the same as 'I eat what I see'!" "You might just as well say," added the March Hare, "that 'I like what I get' is the same thing as 'I get what I like'!" "You might just as well say," added the Dormouse, which seemed to be*

talking in its sleep, "that 'I breathe when I sleep' is the same thing as 'I sleep when I breathe'!"

Alice's statement: "I mean what I say" can be rephrased: "If I say it then I mean it." The Hatter is trying to make the point that this is different from: "I say what I mean.", or "If I mean it then I say it." His example is easier to understand, so we will use that to illustrate the various types of statements:

Type of statement:	IF-Then Statement:	Translation:
Conditional:	*If I eat it then I see it.* **If P then Q.**	I see what I eat.
Converse:	*If I see it then I eat it.* **If Q then P.**	I eat what I see.
Contrapositive:	*If I don't see it then I don't eat it.* **If not Q then not P.**	I don't eat what I don't see.
Inverse:	*If I don't eat it then I don't see it.* **If not P then not Q.**	I don't see what I don't eat.
Negation:	*I eat it and I don't see it.* **P and not Q.**	I don't always see what I eat.

We can now see from this table that the confusion on Alice's part has to do with the **converse** of the conditional. Many people mistakenly think these statements are **logically equivalent**, when they clearly are not as the Hatter's example points out. Eating whatever you see is much different than seeing what you eat!

Two statements are said to be **logically equivalent** if they "mean the same thing." There is a technical definition of this involving truth tables but we will not go into the details, just report two important facts:

1. The **converse** is NOT logically equivalent to its **conditional**.
2. The **contrapositive** is logically equivalent to its conditional.

*As you read, jot down notes and questions. At the end of this section's Guided Worksheets there is a space for **Reflection** and **Monitoring Your Understanding**, where you can try and answer your questions after completing the guided activities and homework.*

Notes & Questions

Guided Practice Activity #1 – Conditionally Speaking

1. State the converse, inverse, contrapositive, and negation of the following statement and indicate a pair of statements that are logically equivalent:

    Conditional: *If they are serving French fries, then I am going to lunch!*

    a. Converse:

    b. Inverse:

    c. Contrapositive:

    d. Negation:

2. Assume the conditional statement is true, and you go to lunch. Does that mean they are serving fries?

3. Which pairs of statements are logically equivalent (there is more than one pair!)?

4. Write the statement "I mean what I say." As a conditional If-then statement.

5. Write the bi-conditional, "I will go to lunch if and only if they are serving fries.", as two conditionals.

6. Which conditional from #5 is the "only if" part?

Section 10.1 Reflections
Monitor Your Understanding

Can you explain the converse error?

Objective 1 – Understand Logic Vocabulary
Objective 2 – Explore Nested IF Functions

Objective 1 – Use IF-THEN-ELSE Syntax

Summary

Conditional statements imply two possible outcomes for a given hypothesis. If it is sunny (the hypothesis) then we go to the beach, else we do not go to the beach. Excel has a built-in **IF function** which allows for two possible outputs, depending on the truth value of the hypothesis. Retailers often offer free shipping if you order more than $99, else it is $5.99. The hypothesis in this example is whether or not the order is over $99. We can create a spreadsheet which has a cell for the size of the order and one for shipping. The first argument of the **IF** function will be the hypothesis, called a *logical test*, because it is either true or false (over $99 or not). In the figure below we have the order in cell **B1** and the shipping in cell **B2**.

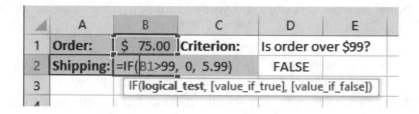

We have entered the logical test in **D2**, =B1>99. Notice that this formula returns the value FALSE because the order is NOT greater than $99. The following table shows the syntax for logical operators used in the logical tests.

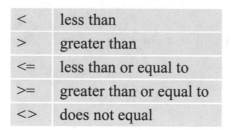

<	less than
>	greater than
<=	less than or equal to
>=	greater than or equal to
<>	does not equal

As you read, jot down notes and questions. At the end of this section's Guided Worksheets there is a space for **Reflection** *and* **Monitoring Your Understanding**, *where you can try and answer your questions after completing the guided activities and homework.*

Notes & Questions

Guided Practice Activity #1 – **IF** Mania

Try out these spreadsheet activities if you dare!

1. Your bank is offering a 1 year Certificate of Deposit (CD) at 1.5% for investments under $5,000. If you invest $5,000 or more they offer 2.2%. What IF function can you enter in cell C5 that gives the proper rate of return?

	A	B	C
1			
2		1 Year Certificate of Deposit	
3			
4		Amount Deposited:	$ 7,500.00
5		Rate of Return:	2.2%
6		Value after 1 year:	$ 7,665.00

2. What is the truth value of the following logical test for the spreadsheet above: = **C4** <= 7500?

3. Your company offers a 401(k) plan for retirement which will match 50% of money you deposit in your 401(k) up to 6% of your total salary (so the maximum match is 3%). The spreadsheet gives your annual salary and your monthly deposit into your 401(k). What formulas are in cells **C5** and **C6** that compute 6% of your salary annually and monthly?

	A	B	C
1			
2		Annual Salary:	$ 50,000.00
3		Monthly Deposit:	$ 400.00
4			
5		Annual 6%:	$ 3,000.00
6		Monthly 6%:	$ 250.00
7			
8		Company Match:	$ 125.00

4. If you deposit $100 a month into your 401(k) what is the company match?

5. If you deposit $789 a month into your 401(k) what is the company match?

6. What **IF** function is entered in cell **C8** that computes the company match?

Summary

You run a business making candles and offer customers a 20% discount on their order if they order twenty or more candles. We create a spreadsheet which has a cell for the logical test, Order >= 20, and uses the **IF** function to compute the total cost of the order if a single candle costs $2.50.

B2	▼	⋮	✕	✓	f_x	=IF(B1>=20, 80%*2.5, 2.5)		
	A		B		C	D	E	F
1	Order:		22		Criterion:	Is order 20 or more candles?		
2	Price per candle:	$	2.00			TRUE		
3	Total cost:	$	44.00					

Note we could have just computed the total cost, using the formula,

$$\text{=IF(B1>=20, 80\%*2.5*B1, 2.5*B1)}$$

but chose to compute price per candle first to break the problem into simpler steps. The previous example illustrated using the **IF** function to handle two possible outputs, and using a tree diagram can help keeping track of the outputs.

FIGURE 10.5 Candle Order Tree Diagram

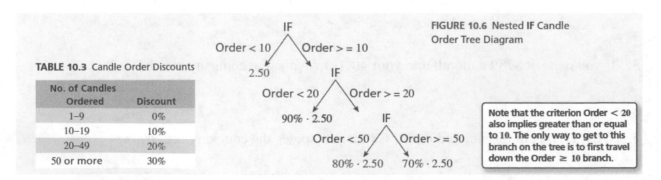

We can handle more than two outputs using nested **IF** functions. Your candle company has 4 possible discounts based on the number candles ordered as shown in the table. Our tree diagram is now more complicated:

TABLE 10.3 Candle Order Discounts

No. of Candles Ordered	Discount
1–9	0%
10–19	10%
20–49	20%
50 or more	30%

FIGURE 10.6 Nested IF Candle Order Tree Diagram

Note that the criterion Order < 20 also implies greater than or equal to 10. The only way to get to this branch on the tree is to first travel down the Order ≥ 10 branch.

The figure above illustrates out why tree diagrams can assist in tracking all of the possible outputs. Remember that the right branch represents the ELSE option in the IF-THEN-ELSE format; with more than two possible outputs this third ELSE option becomes another **IF function** as shown in **Figure 10.7** below.

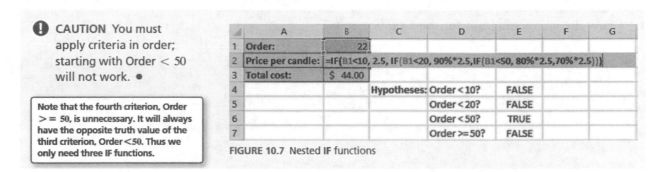

> ❗ **CAUTION** You must apply criteria in order; starting with Order < 50 will not work. •

> Note that the fourth criterion, Order >= 50, is unnecessary. It will always have the opposite truth value of the third criterion, Order <50. Thus we only need three IF functions.

	A	B	C	D	E	F	G
1	Order:	22					
2	Price per candle:	=IF(B1<10, 2.5, IF(B1<20, 90%*2.5,IF(B1<50, 80%*2.5,70%*2.5)))					
3	Total cost:	$ 44.00					
4			Hypotheses:	Order <10?	FALSE		
5				Order <20?	FALSE		
6				Order <50?	TRUE		
7				Order >=50?	FALSE		

FIGURE 10.7 Nested IF functions

As you read, jot down notes and questions. At the end of this section's Guided Worksheets there is a space for **Reflection** *and* **Monitoring Your Understanding***, where you can try and answer your questions after completing the guided activities and homework.*

Notes & Questions

Guided Practice Activity #1 – Zip Shipping

In this example there are two possibilities for the shipping amount. **IF** the order is < $100, **THEN** shipping is $7.99 **ELSE** it is free!

	A	B	C	D	E
1	Zip Shipping				
2	Input Order:	$ 132.56			
3	Output shipping fee:	=IF(			
4		IF(**logical_test**, [value_if_true], [value_if_false])			

1. What IF function is typed into cell **B3**?

2. What function would you type in if shipping is 10% of the order, for orders less than $100, and 5% of the order for all other orders?

3. We can also handle situations with more than two possibilities by *nesting* **IF** functions. Assume there are three possibilities: 10% for < $100, 5% for < $250, and free for all others. Now what is the IF function?

4. Draw the arrow diagram for the situation in **#3**.

Excel Preview – 10.1 Overtime

You run a company with 6 employees. Create a spreadsheet that keeps track of the hours each employee works and determines their weekly salary. If they work 40 hours or less they simply get their hourly wage times the number of hours worked. If they work overtime, anything over 40 hours gets paid time and a half; so 2 overtime hours gets paid as 3 hours worked.

	A	B	C	D	E	F	G	H	I	J	K	L
7	Employee				Daily Hours Worked				a.			
8	Last	First	Mon	Tue	Wed	Thu	Fri	Hourly Wage	Total Hours		Hours Over 40	Weekly Salary
9	Ackers	Jill	4	4	4	4	4	$ 12.50	20		0	$ 250.00
10	Bartel	Channing	11	10	4	5	10	$ 10.00	40		0	$ 400.00
11	First	Last	9	7.5	6.5	7	13	$ 10.00	43		3	$ 445.00
12	Hill	Jonah	2	1	0	1.5	0.1	$ 2.25	4.6		0	$ 10.35
13	Jackson	Five	12	8	14	22	16	$ 45.00	72		32	$ 3,960.00
14	Yu	Tu	5	8	7	8	7	$ 150.00	35		0	$ 5,250.00
15												

1. Assume you make $10 per hour and work 42 hours. What is your weekly salary?

2. What **IF** function can you enter in cell **K9** that can be filled down?

3. What **IF** function can you enter in cell **L9** that can be filled down?

Section 10.2 Reflections
Monitor Your Understanding

Can you explain nested **IF** functions to someone else?

Objective 1 – Understand VLOOKUP Syntax

Summary

Recall the candle order example:

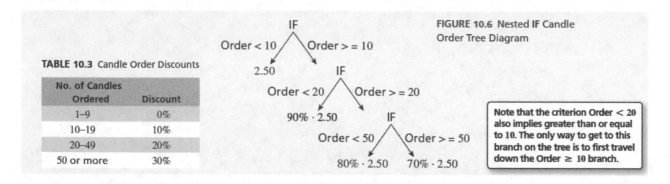

TABLE 10.3 Candle Order Discounts

No. of Candles Ordered	Discount
1–9	0%
10–19	10%
20–49	20%
50 or more	30%

FIGURE 10.6 Nested IF Candle Order Tree Diagram

Note that the criterion Order < 20 also implies greater than or equal to 10. The only way to get to this branch on the tree is to first travel down the Order ≥ 10 branch.

Instead of using nested **IF** functions it is much more efficient to use the following table and the **VLOOKUP** function.

No. Candles Ordered	Left Hand Endpoints	Discount
1-9	1	0%
10-19	10	10%
20-49	20	20%
50 or more	50	30%

We can enter this table into Excel and then use **VLOOKUP** which has three arguments:

=**VLOOKUP**(Lookup this Value,   in this Table,   get Output from this Column)

The first two arguments as shown in Figure 10.8 below are self-explanatory: We want to look up the size of the order in the table in cells **A7:B10**. The order is an input cell and any value may be input in cell **B1**. The third argument, *col_index_num*, requires care; it is the number of the column that you want for your output from the **VLOOKUP** function. In this case we want to find the discount factor in *Column #2* so we put two in for the third argument.

Note that we can multiply the **VLOOKUP** function by 2.5 in a formula to get the price per candle.

CAUTION Excel always looks in the first column of the table for the lookup value. Remember that 10 in the first column stands for the range 10–19. ●

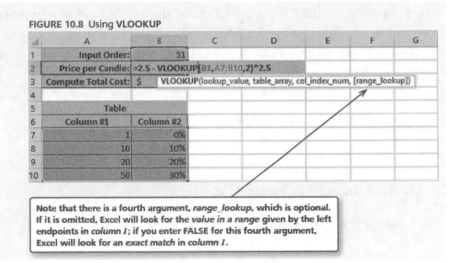

FIGURE 10.8 Using VLOOKUP

	A	B	C	D	E	F	G
1	Input Order:	51					
2	Price per Candle:	=2.5 - VLOOKUP(B1,A7:B10,2)*2.5					
3	Compute Total Cost:	$ VLOOKUP(lookup_value, table_array, col_index_num, [range_lookup])					
4							
5	Table						
6	Column #1	Column #2					
7	1	0%					
8	10	10%					
9	20	20%					
10	50	30%					

Note that there is a fourth argument, *range_lookup*, which is optional. If it is omitted, Excel will look for the *value in a range* given by the left endpoints in *column 1*; if you enter FALSE for this fourth argument, Excel will look for an *exact match* in *column 1*.

*As you read, jot down notes and questions. At the end of this section's Guided Worksheets there is a space for **Reflection** and **Monitoring Your Understanding**, where you can try and answer your questions after completing the guided activities and homework.*

Notes & Questions

Guided Practice Activity #1 – Zip Shipping Nested

If there are many possibilities the nesting of **IF** functions gets out of hand!

	A	B	C
1	Zip Shipping		
2	Bins	Cutoffs	Output %
3	$0-$99.99	$ -	10.00%
4	$100-$249.99	$100.00	5.00%
5	$250-$499.99	$250.00	4.00%
6	$500-$999.99	$500.00	2.50%
7	$1000-	$1,000.00	0.00%
8			
9	Input Order Total:	$874.00	
10	Shipping Percentage:	2.50%	
11	Shipping costs:	$21.85	
12			

In this situation we have 5 possibilities for the %. Instead of using nested **IF** functions we represent the possibilities in a table in cells, **B3:C7**. This table consists of two columns, #1 and #2. Basically we will LOOKUP the value for the order total **B9**, in the table, and then output the appropriate % from column # 2.

1. What **VLOOKUP** function can you enter in cell **B10**?

2. What are the outputs for the following?

   a. =**VLOOKUP**(**165**, B3:C7, 2)

   b. =**VLOOKUP**(650, B3:C7, **1**)

   c. =**VLOOKUP**(650, B3:C7, **2**)

   d. =**VLOOKUP**(6500, B3:**C6**, 2)

Summary

There may be times when you wish to test for multiple criteria at once. To do this Excel has additional logic functions: **AND**, **OR**, and **NOT**, which can be used to make compound logical tests. The **AND** function is true if and only if all of the arguments are true, while the **OR** function is true if and only if at least one of the arguments is true.

Examples:

- =**AND**(a cat is an animal, trees inhale CO2, owls are birds)… TRUE
- =**AND**(a cat is an animal, trees walk around, owls are birds)… FALSE
- =**OR**(a group of kittens is called a kindle, trees like ice cream, owls don't poop)… TRUE

Well, that was fun and I made you type "poop" into a search engine. There are serious applications of these functions as explored in the next activity.

As you read, jot down notes and questions. At the end of this section's Guided Worksheets there is a space for ***Reflection*** *and* ***Monitoring Your Understanding****, where you can try and answer your questions after completing the guided activities and homework.*

Notes & Questions

Guided Practice Activity #1 – Expedia

Expedia is offering a deal, 20% off your purchase if you book airfare and a hotel or book airfare and rent a car. Create a spreadsheet which allows you to enter the price of airfare, hotel, and car rental. If airfare and hotel are both greater than zero, or airfare and car are both greater than zero, take 20% off of the whole package.

	A	B	C	D
1	Costs		Logical Tests	
2	Hotel:	$ 480.00	Hotel&Air:	TRUE
3	Airfare:	$ 555.00	Car&Air:	FALSE
4	Car:	$ -	D2 OR D3:	TRUE
5	Discount %:	20%		
6	SubTotal:	$1,035.00		
7	Discount:	$ 207.00		
8	Total:	$ 828.00		

1. What logical function is typed into cell **D2** that tests if there is a cost entered for hotel and for airfare?

2. What function is typed into cell **D4** that tests if at least one of **D2** or **D3** is true?

3. What function is typed into cell **B5** to determine if there is a 20% discount?

4. What is the output of the following functions?

   a. =NOT(AND(B2<500, B8<>800))

   b. =OR(B5<1, B6>10^5, B7>500)

   c. =AND(SUM(B2:B5)>1000, MIN(B5:B8)<15/100)

Excel Preview – 10.2 Credits

This spreadsheet has an input cell for the number of credits earned by a student, which determines if the student is a First-year (<= 32), sophomore (<= 64), junior (<= 96), or senior (> 96). Recall that "text" in an **IF** function needs to be in quotation marks.

	A	B	C	D	E
1			Last	First	
2		Student:	Azura	Bella	
3		Credits:	47		
4		a.) Year:	Sophomore	using the IF function	
5					
6				VLOOKUP Table	
7			Intervals for Credits	Left Hand Endpoints	Class Year
8			0.. 32		First-year
9			33.. 64		Sophomore
10			65.. 96		Junior
11			97..		Senior
12					
13		c.) Year:	#N/A	using the VLOOKUP function	

1. What IF function is entered in cell **C4**?

2. What is entered in cells **D8:D11**?

3. What VLOOKUP function is entered in cell **C13**? The current output in the screenshot is giving the **#N/A** error because cells **D8:D11** are blank.

4. Which do you like better, **IF** or **VLOOKUP**, here?

Section 10.3 Reflections
Monitor Your Understanding

What is the difference between **IF** and **VLOOKUP**?

Section 10.4 Random Simulations

Objective 1 – Use RAND and RANDBETWEEN
Objective 2 – Conduct Random Simulations

Objective 1 – Use RAND and RANDBETWEEN

Summary

The Logic **functions** studied in this chapter open up a new range of problems which have more than one output. We will finish the chapter with a powerful application of these new tools, modeling real world phenomena using random simulations. The idea will be to use a random number generator, either **RAND** or **RANDBETWEEN**, and **VLOOKUP** to randomly choose possibilities such as possible stock returns or gas prices. First, we must introduce the **RAND** function which generates a random number between zero and one:

$$0 < \textbf{RAND}(\,) < 1$$

CAUTION The **RAND** function does not have any arguments but requires the open and close parentheses. ●

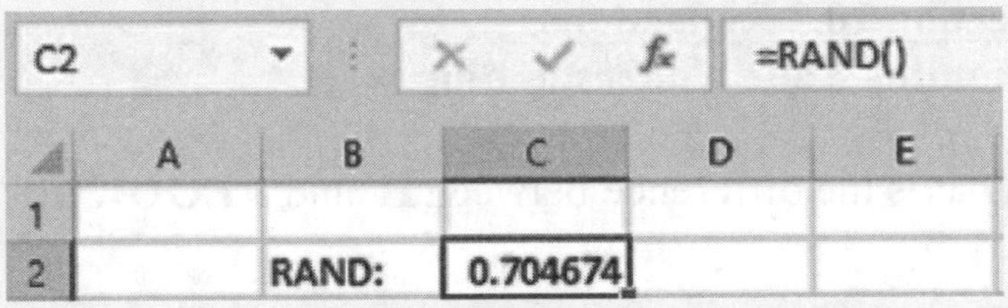

FIGURE 10.16 **RAND** function

The **RANDBETWEEN** function will return a random number between the two arguments:

$$2 <= \textbf{RANDBETWEEN}(2, 17) <= 17$$

*As you read, jot down notes and questions. At the end of this section's Guided Worksheets there is a space for **Reflection** and **Monitoring Your Understanding**, where you can try and answer your questions after completing the guided activities and homework.*

Notes & Questions

Guided Practice Activity #1 – Random Fun

We will study the **RAND** function, which generates a random number between 0 and 1. Typing in =**RAND()** results in a random number between 0 and 1, for example: 0.527834251. Note that **RAND** has no arguments but the parentheses are required, and that **IF** requires quotation marks around text in the second and third arguments.

1. What does the following **IF** function do: = IF(RAND() < 0.5, "Heads", "Tails") ?

2. Write down an **IF** function which flips an unfair coin with heads coming up 75% of the time.

3. Write down an **IF** function that returns the number 1, a third of the time, the number 2, a third of the time and the number 3, a third of the time.

4. Create a spreadsheet that simulates rolling a die using the **VLOOKUP** and **RAND** functions. Write down all the formulas in cells **B3:C8**, and **C10**.

	A	B	C	D
1	Random #:	0.207697		
2		TABLE		
3				
4				
5				
6				
7				
8				
9				
10	Die Roll:			

Summary

We can use RAND and RANDBETWEEN to conduct random simulations. The text contains a detailed example. Here we provide a simpler activity for exploration. Enjoy!

As you read, jot down notes and questions. At the end of this section's Guided Worksheets there is a space for **Reflection** *and* **Monitoring Your Understanding***, where you can try and answer your questions after completing the guided activities and homework.*

Notes & Questions

Guided Practice Activity #1 – IPO Investing

Your friend who graduated with an underwater synchronized dance degree just landed a job at a hedge fund (answering phones) where they overheard a hot stock tip. By investing in ultra-risky initial public offerings (IPO's) each week you either can earn 80% or lose 60% (completely random, 50-50 chance of which occurs), which means you should average a 10% gain! Right?! RIGHT?!

1. Assuming you invest $10,000 and then average a 10% gain each week for 52 weeks, compute how much your investment is now worth.

2. Consider the spreadsheet shown below which randomly assigns one of the outcomes, 80% or -60%, for each week in 1 year using the **IF** function and the **RAND** function:
   IF RAND()<0.5 THEN 80% ELSE -60%.

   i. What formula is entered in cell **C2**?

   ii. What formula is entered in cell **D2**?

   iii. What formula is entered in cell **B3**?

iv. What formula is entered in cell **F2**?

	A	B	C	D	E	F	G	H
1	Week	Balance	Return	End Balance		Balance at the end of the year:		
2	1	$ 10,000.00	-60%	$ 4,000.00		$ 1.95		
3	2	$ 4,000.00	-60%	$ 1,600.00				
4	3	$ 1,600.00	80%	$ 2,880.00				
5	4	$ 2,880.00	-60%	$ 1,152.00				
6	5	$ 1,152.00	-60%	$ 460.80				
7	6	$ 460.80	-60%	$ 184.32				
8	7	$ 184.32	80%	$ 331.78				
9	8	$ 331.78	80%	$ 597.20				
10	9	$ 597.20	80%	$ 1,074.95				
11	10	$ 1,074.95	-60%	$ 429.98				
12	11	$ 429.98	80%	$ 773.97				
13	12	$ 773.97	-60%	$ 309.59				

3. How does the end of year balance shown compare to your result in **part 1**?

4. If we keep track of the ending balance for 10 different scenarios we get the results shown below. Average these 10 ending balances. Note that if you run the simulation, your ending balances will be different and you could have some that are very large. That is the point of a random simulation, it's random!

    v.   Scenario 1: $1.95
    vi.   Scenario 2: $0.00
    vii.   Scenario 3: $177.94
    viii.   Scenario 4: $1.95
    ix.   Scenario 5: $800.75
    x.   Scenario 6: $0.10
    xi.   Scenario 7: $0.00
    xii.   Scenario 8: $1.95
    xiii.   Scenario 9: $800.75
    xiv.   Scenario 10: $8.79

5. Is this average close to what you computed in **#1** of this problem? Why not!?

6. Compute the geometric mean of 80% and -60%. Use this average rate of return to re-compute the worth of your investment after 52 weeks. Is this closer to your average of 10 ending balances?

Excel Preview – 10.9 Investing 2

This spreadsheet simulates an investment with variable annual rates of return for 10 years: a 25% chance of losing 5%, a 55% chance of returning 4%, a 15% chance of returning 10%, and a 5% chance of making 20%.

Assume you start with no money in the account but make an initial deposit of $5,000.

- Enter a formula for the variable rate of return by looking up a random number in the table.
- Compute the ending balance by multiplying the sum of the balance and deposit by the growth factor associated with the growth rate.
- Connect the starting balance of year 2 to the ending balance of year 1.
- Make the deposit grow by 4% in year 2. Fill down the formulas for rate of return and ending balance for year 2. Then fill down all of row 2 to year 10.

	A	B	C	D	E	F	G	H
1								
2		a.)	a.)	a.)	a.)			
3	Year	Balance	Deposit	Rate of Return	Ending Balance		Cutoff	Rate of Return
4	1	$ -	$ 5,000.00	-5%	$ 4,750.00		0	-5%
5	2	$ 4,750.00	$ 5,200.00	10%	$ 10,945.00		25%	4%
6	3	$ 10,945.00	$ 5,408.00	4%	$ 17,007.12		80%	10%
7	4	$ 17,007.12	$ 5,624.32	-5%	$ 21,499.87		95%	20%
8	5	$ 21,499.87	$ 5,849.29	4%	$ 28,443.13			
9	6	$ 28,443.13	$ 6,083.26	4%	$ 35,907.45			
10	7	$ 35,907.45	$ 6,326.60	-5%	$ 40,122.34			
11	8	$ 40,122.34	$ 6,579.66	-5%	$ 44,366.90			
12	9	$ 44,366.90	$ 6,842.85	4%	$ 53,258.13			
13	10	$ 53,258.13	$ 7,116.56	4%	$ 62,789.68			

1. What is the chance of getting a 10% rate of return?

2. If you wanted a 40% chance of returning 4% and a 30% chance of returning 10%, what should you change the cutoff 80% to?

3. What formula is in cell **D4** that can be filled down?

4. What formula is in cell **E5** that can be filled down?

Section 10.4 Reflections
Monitor Your Understanding

What does it mean to conduct a random simulation?

Chapter 11 – Chance, Risk, and Likelihood

Section 11.1 Risk and Decision Making

Objective 1 – Understand Uncertainty

Key Terms

Certainty/chance/risk/ Likelihood	

Summary

This text, *Thinking Quantitatively*, is dedicated to providing you with the quantitative skills needed for informed decision making in your personal, professional, and public life. Often we are faced with situations that involve **chance**, or have incomplete information resulting in **likelihoods** and **risk** assessment. To deal with uncertainty we need a firm grasp of **probability**, which is the focus of this chapter. It is important to understand that science does not provide definitive proof, just increasing certainties based on available evidence and experimental verification.

To make an informed decision as a citizen regarding public policy issues one must understand that life is filled with uncertainty. Smoking does not always lead to lung cancer and scientists cannot predict exactly how much temperatures will rise given the best available data on greenhouse gas emissions. We also need to appreciate that **chance** is not absolute but relative to other factors. The chance of a 55 year old person dying from a heart attack is highly dependent on whether they smoke. The chance of dying from a gunshot is 30 times higher in the U.S. than in England.

TABLE 11.2 Comparison of Death Rates (2007–2012 Averages)

Being killed with a gun in:	Is as likely as dying from ____ in the U.S.	Deaths per million
El Salvador	Heart attack	446.3
United States	Car accident	31.2
China	Plane crash	1.6
Japan	Lightning strike	0.1

Data from: http://www.nytimes.com/2015/12/05/upshot/in-other-countries-youre-as-likely-to-be-killed-by-a-falling-object-as-a-gun.html?_r=0

Table 11.2 compares the risks of dying from a gunshot in other countries with the U.S., and shows living in the United States makes you much more likely to be a gun homicide victim. Living in Chicago in 2015 raises the rate of gun deaths to 178 per million, while being poor, black, male and living in Chicago raises the rate even more.

In 2015 gun control and race relations were topics that dominated the headlines and polarized the debates of our Presidential hopefuls. Should we vote for stricter gun control measures? Just how dangerous is our society? Does racial profiling impact police tactics such as stop and frisk? Answering such questions requires understanding **risk** and assessing various interventions. We can reduce our risk of dying from a car crash to almost zero by walking everywhere and similarly reduce our risk of a heart attack, stroke, or diabetes by adopting a whole food plant based diet. This chapter will provide you with the quantitative skills to help make such decisions.

As you read, jot down notes and questions. At the end of this section's Guided Worksheets there is a space for **Reflection** *and* **Monitoring Your Understanding***, where you can try and answer your questions after completing the guided activities and homework.*

Notes & Questions

Guided Practice Activity #1 – Know Your Chances

The table below gives the risk of dying from various diseases for 55 year old men and women by smoking status. These numbers are rates: deaths per 1,000 of each cohort over the next 10 years.

Table 11.1: Death Rates: Numbers gives deaths per 1,000 of each cohort over the next 10 years

Age	Sex	Smoking Status	Vascular Disease		Cancer					Lung Disease		
			Heart Attack	Stroke	Lung	Breast	Colon	Prostate	Ovarian	COPD	Accidents	All Causes
55	M	Never Smoked	19	3	1		3	2		1	5	74
55	M	Smoker	41	7	34		3	1		7	4	178
55	F	Never Smoked	8	2	2	6	2		2	1	2	55
55	F	Smoker	20	6	26	5	2		2	9	2	110

Data From: Steve Woloshin Lisa Schwartz, and H. Gilbert Welch, "The Risk of Death by Age, Sex, and Smoking Status in the U.S.: Putting Health Risks in Context," *Journal of the National Cancer Institute* 100 (2008): 845-853[1]

Is smoking bad for you? How do you know?

1. Who is more likely to die in the next 10 years from a heart attack: a 55 year old man or woman?

2. By what factor does smoking increase the chance of a 55 year old man/woman dying from lung cancer?

3. Does smoking cause heart attacks or minimize the chance of getting breast cancer?

---

[1] http://www.ncbi.nlm.nih.gov/pmc/articles/PMC3298961/

4. Should you smoke?

Does human activity influence the climate? How do you know?

The Intergovernmental Panel on Climate Change (IPCC) issued their fifth assessment report (AR5[2]) in 2014 and concluded that human greenhouse gas emissions "have been detected throughout the climate system and are *extremely likely* (95-100% certain) to have been the dominant cause of the observed warming since the mid-twentieth century." Note that the figure below shows observed data in line with scientists' predictions. Surface temperature is projected to rise over the 21st century under all assessed emission scenarios. It is *very likely* (90-100% certain) that heat waves will occur more often and last longer, and that extreme precipitation events will become more intense and frequent in many regions. The ocean will continue to warm and acidify and global mean sea level to rise (change relative to 1986-2005 average).

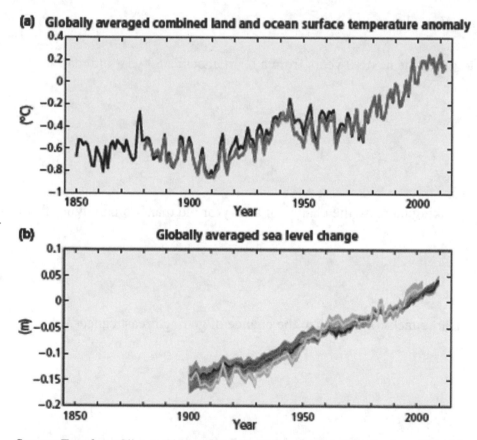

Source: Text from Climate Change 2014: Synthesis Report. Contribution of Working Groups I, II, and III to the Fifth Assessment Report of the IPCC. [Core Writing Team, Pachauri, R.K. and Meyer, L. 9eds.)], pp 3 IPCC, Geneva, Switzerland.

---

[2] https://www.ipcc.ch/assessment-report/ar5/

5. In **Figure (a)** above interpret the value for 1850 and 2010.

6. In **Figure (b)** the data looks linear. Sketch in a straight line and find the equation.

7. Interpret the slope and y-intercept of your line.

8. In **Figure (c)** below there are 3 trendlines for CO₂ (top line), CH₄ (middle), and N₂O (bottom). There are different axes and units for all three! Which of the three molecules is most concentrated in the atmosphere? Recall a CO₂ concentration of 400 ppm means 400 parts per million, indicating for every 1 million molecules of air, 400 of these are CO₂. Also ppb is parts per billion.

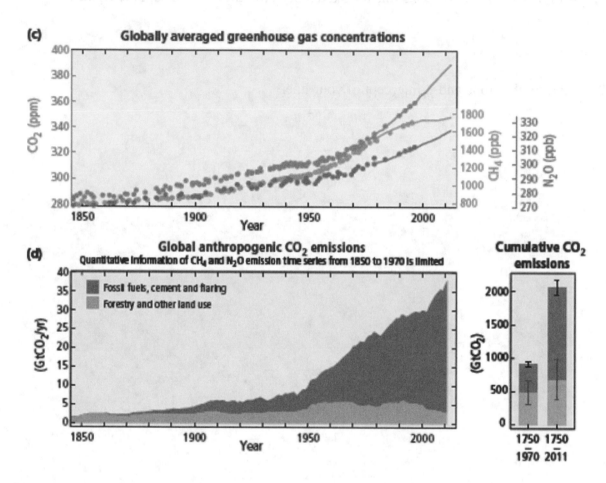

9. What percentage of the atmosphere is made up of each of the three molecules in 2010?

10. Given these are such small percentages, why is there cause for concern about the indicated concentrations?

11. In **Figure (d)** how does the graphic on the left differ from the cumulative emissions graphic on the right?

12. There is uncertainty in the cumulative CO2 emissions from 1750 to 2011. What do scientists say are the smallest and largest cumulative CO2 emissions over this time period?

13. What is the estimated cumulative CO_2 emissions from 1970 to 2011?

14. Find the average cumulative CO_2 emissions per year over the two different time periods: 1750-1970 and 1970-2011.

15. We have seen above how bad smoking is for us. Does it seem reasonable to believe that human caused greenhouse gas emissions have no effect on the climate system?

Excel Preview – 11.4 Casino

This section deals with chance and risk in general. This Excel problem asks you to use the logic function from Chapter 10 to create a simulation called *Gambler's Ruin*. It is a good lesson in the power of modeling to generate insights for informed decision making. You are going to flip a coin and bet on Tails coming up. You start with a $10 bet and will double your bet if you lose, once you win you revert to the $10 bet. **This guarantees you a profit!!!**

◢	A	B	C	D	E
1		Bet	Coin Flip	# H in a Row	Winnings
2	1	$ 10.00	T	0	$ 10.00
3	2	$ 10.00	T	0	$ 20.00
4	3	$ 10.00	H	1	$ 10.00
5	4	$ 20.00	T	0	$ 30.00
6	5	$ 10.00	H	1	$ 20.00
7	6	$ 20.00	H	2	$ -
8	7	$ 40.00	H	3	$ (40.00)
9	8	$ 80.00	H	4	$ (120.00)
10	9	$ 160.00	T	0	$ 40.00
11	10	$ 10.00	H	1	$ 30.00

1. Explain what happens in the second row of the simulation, i.e. what are these outputs in cells **B3:E3**?

2. Why is the bet $20 in cell **B5**?

3. Explain how the winnings got to be -$120 in cell E9.

4. What is the probability of getting 4 heads in a row? Try to determine this probability, we will develop the mathematics behind the solution in later sections.

5. Explain why every time we win after a losing streak (heads in a row) the winnings are $10 more than our previous winnings before the losing streak.

6. The bet is a basic **IF** function that doubles our bet if we lose. What function can we enter in cell **B3** that can be filled down?

7. The winnings is also a basic **IF** function that adds or subtracts the bet. What function can we enter in cell **E3** that can be filled down?

8. This one is more difficult, how can we count the number of heads in a row? What function can we enter in cell **D3** that can be filled down?

9. Why is this called "Gambler's Ruin" if we are guaranteed to continue adding $10 to our winnings?

Section 11.1 Reflections
Monitor Your Understanding

What surprised you in this section?

Objective 1 – Explore Events, Outcomes, and Sample Spaces
Objective 2 – Understand Permutations and Combinations

Objective 1 – Explore Events, Outcomes, and Sample Spaces

Key Terms

Probability	
Event Outcome Sample space	
Odds	

Summary

The **probability** of an **event** occurring is a measure of the **likelihood** or **chance** of it to occur. This idea of quantifying the uncertain is inherently un-intuitive. Indeed human beings have been gambling, and playing games of chance since before recorded history yet mathematicians did not formally "discover" the laws of probability until the 17th century. This is probably because they viewed "chance" or "fate" as coming from divine intervention; the Romans worshipped the goddess Fortuna of luck and chance, and ancient oracles used dices and bones to divine fates pre-determined by the gods (the I-Ching is still very popular today). If chance is due to the caprice of the gods, why bother trying to quantify it? The mathematics of probability is not terribly sophisticated, just counting and **ratios**.

We will start with a simple example to illustrate the concepts and vocabulary. You meet with a friend who has two children. What is the probability her second child is a girl? She mentions one of her children is a boy. Given this new information what is the probability that her second child is a girl? Assume the probability of having a boy or girl is equally likely (50%).
The **probability** of the **event**, *2nd child girl*, is the **ratio** of the number of ways this event can occur to all possible **outcomes** of having 2 children. We can create a table showing the collection of all possible outcomes (called the **sample space**):

	Second Child	
First Child	Boy	Girl
Boy	**BB**	**BG**
Girl	**GB**	**GG**

So, the sample space consists of 4 possible outcomes: {BB, GB, BG, GG}. The event, *2nd child girl*, consists of 2 possible outcomes: {BG, GG}. Thus the probability of the 2nd child being a girl is 2 : 4 or 50%.

If we know that one of the children is a boy, our sample space changes: {BB, GB, BG}; so that the event now consists of a single element: {BG}, and the probability is 1 : 3 or 33.3%. As we saw in the previous section, additional information can radically change the **likelihood** of an event!

As you read, jot down notes and questions. At the end of this section's Guided Worksheets there is a space for **Reflection** *and* **Monitoring Your Understanding***, where you can try and answer your questions after completing the guided activities and homework.*

Notes & Questions

Guided Practice Activity #1 – Know Your Children

You meet a friend who has three children.

1. What is the sample space for all the possible ways she can have three children?

2. List the outcomes of the event, *3rd child girl*.

3. What is the probability her third child is a girl, **P** (*3rd child girl*)?

4. She mentions one of her children is a boy. What is the new sample space?

5. What is the new probability her third child is a girl, **P** (*3rd child girl*)?

6. If she then mentions she has two girls (and also a boy), what is the new probability that her third child is a girl?

Key Terms

Fundamental principle of counting	
Permutation	
Combination	

Summary

Counting is critical to computing **probabilities**, and the previous guided activity listing the possible ways to have 3 children, {BBB,BBG,BGB,BGG,GBB,GBG,GGB,GGG}, illustrates what is called the **Fundamental Principle of Counting**. We have 3 children, or 3 choices to be made, and 2 options for each child/choice so we multiply the number of options for each choice: $2 \times 2 \times 2 = 8$. We can illustrate this with a tree diagram:

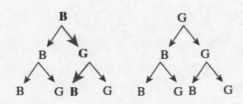

FIGURE 11.4 Tree Diagram Listing Number of Ways to Have Three Children

Each branch of the tree represents a possible way to have 3 children, the highlighted branch gives BGB. There are 8 such branches: we have 2 options for the 1st child and for each of these there are 2 options for the 2nd child giving 4 ways to have 2 children. For each of these 4 possibilities there are 2 options for the 3rd child giving 8 total possibilities for 3 children.

Permutations and combinations are used for counting different events.

Given a group of 6 people: {Ann, Mei, Suk, Bob, Jim, Rav}, how many ways can these 6 people finish a race 1st, 2nd, and 3rd? How many ways can we choose a committee of 3?

To finish a race, order does matter. We have 6 choices for 1st, then only 5 choices for 2nd and 4 choices for 3rd, giving us $6 \times 5 \times 4 = 120$ ways to finish the race or 120 permutations, $_6P_3$:

To choose committees, order does NOT matter so we arrange our 120 permutations into groups of 6: {abc, acb, bac, bca, cab, cba} each of which represents a single committee. Thus the number of committees is the number of combinations, $_6C_3 = \dfrac{_6P_3}{3!} = \dfrac{120}{6} = 20$ committees.

As you read, jot down notes and questions. At the end of this section's Guided Worksheets there is a space for **Reflection** *and* **Monitoring Your Understanding**, *where you can try and answer your questions after completing the guided activities and homework.*

Notes & Questions

Guided Practice Activity #1 – Counting

1. You have 2 shirts, 3 shorts, and 3 hats to choose from to make an outfit. How many different outfits are possible?

2. Draw a tree diagram illustrating the possibilities.

3. Given 12 people in a race, how many ways can they finish first and second?

4. Given 12 people in a class, how many ways can we choose a committee of 4?

5. There are 12 people in your class and a group of 4 students are randomly being chosen to attend a symposium. What is the probability you and your best friend in the class will both be chosen to go?

   a. First determine how many ways we can choose the other two students.

   b. Then compute the probability.

Excel Preview – 11.1 Coin Flips

In the chapter we saw an example showing how to use Excel to flip 2 coins and count the number of times you get exactly 1 head. We will reconstruct this spreadsheet and compare flipping 2 coins 10 times, 100 times and 1,000 times.

	A	B	C	D	E	F	G	H	I	J	K	L	M	N
1	In the chapter we saw an example showing how to use Excel to flip 2 coins and count the number of times you get exactly 1 head. W													
2														
3														
4		a.)	a.)	b.)			a.)	a.)	b.)			a.)	a.)	b.)
5		Flipping Coins 1,000 times					Flipping Coins 100 times					Flipping Coins 10 times		
6		1st Coin	2nd Coin	# Heads			1st Coin	2nd Coin	# Heads			1st Coin	2nd Coin	# Heads
7	1	T	T	0		1	H	T	1		1	H	H	2
8	2	H	H	2		2	T	T	0		2	H	H	2
9	3	H	H	2		3	H	T	1		3	T	T	0
10	4	H	T	1		4	H	T	1		4	H	H	2
11	5	T	H	1		5	T	T	0		5	H	H	2
12	6	H	H	2		6	T	T	0		6	H	H	2
13	7	H	T	1		7	T	H	1		7	T	T	0
14	8	T	T	0		8	H	H	2		8	H	T	1
15	9	T	H	1		9	H	T	1		9	T	H	1
16	10	H	H	2		10	H	T	1		10	T	T	0
17	11	T	H	1		11	T	H	1					

1. What function is entered in cell **B7** that can be filled own?

2. What function is entered in cell **D7** that can be filled own?

3. What is the theoretical probability of getting exactly one head when flipping two coins?

4. What is wrong with the following argument?

    If you flip two coins, there are three outcomes, $\{0H, 1H, 2H\}$, so $\mathbf{P}(1H) = \dfrac{1}{3}$.

5. How are we going to be able to compute the empirical probability of getting exactly one head using this simulation?

6. Which of the three situations (10 flips, 100 flips, or 1000 flips) will give an empirical probability of exactly one head consistently closest to the theoretical probability of $\dfrac{1}{2}$?

Section 11.2 Reflections
Monitor Your Understanding

What results surprised you in this section?

Objective 1 – Understand Conditional Probabilities
Objective 2 – Explore the First Law of Probability

Objective 1 – Understand Conditional Probabilities

Key Terms

Conditional probability	
Set	
Subset	
Intersection	
Disjoint	
Venn diagram	

Summary

At the end of section 11.1, we discussed how **chance** was highly dependent on other factors or conditions such as where you live and lifestyle choices such as smoking. **Probabilities** that are based on a given condition are called conditional probabilities.

Recall the **two way table** from section 4.4 involving the use of a skin cream given in **Table 11.5**. We can compute the **conditional probabilities**:

$$\mathbf{P}\left(Got\ Better | Used\ Cream\right) \text{ and } \mathbf{P}\left(Used\ Cream | Got\ Better\right).$$

	Rash got Better	Rash got Worse	Totals
Did use cream	223	75	**298**
Did not use cream	107	21	**128**
Totals	**330**	**96**	

We can see from the table that 223 people satisfy both criteria, (*Got Better and Used Cream*), and 298 people total used the cream, so the probability that you got better given you used the cream: $\mathbf{P}\left(Got\ Better|Used\ Cream\right) = \dfrac{223}{298} = 74.8\%$. We can turn the table into a Venn diagram in Figure 11.8 and see that out of the 330 people who got better, 223 used the cream, so the probability that you used the cream given you got better:

$$\mathbf{P}\left(Used\ Cream|Got\ Better\right) = \dfrac{223}{330} = 67.6\%.$$

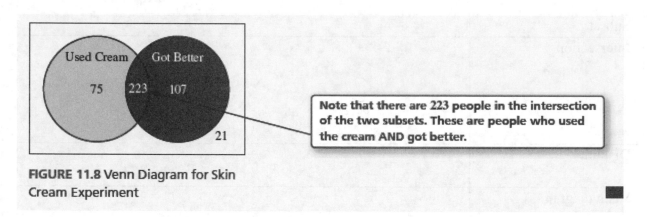

Note that there are 223 people in the intersection of the two subsets. These are people who used the cream AND got better.

FIGURE 11.8 Venn Diagram for Skin Cream Experiment

CAUTION! The conditional probabilities have a **sample space** (denominator) restricted by the given condition. Note that switching the order of the event and the condition changes the probability.

As you read, jot down notes and questions. At the end of this section's Guided Worksheets there is a space for **Reflection** *and* **Monitoring Your Understanding**, *where you can try and answer your questions after completing the guided activities and homework.*

Notes & Questions

Guided Practice Activity #1 – On What Condition

Given the following table:

	Male	Female	Totals
Smoker	130	75	**205**
Non-smoker	250	215	**465**
Totals	**380**	**290**	

1. Use the number 75 in a sentence.

2. What is the greater, the percentage of women who smoke or the percentage of smokers who are women?

3. Compute the probabilities: $P(Smoker|Male)$ and $P(Female|Non\text{-}smoker)$.

4. Turn the table into a Venn Diagram.

5. If choose one person at random from this group what is the probability they are a female non-smoker?

6. Given someone from this group is female what is the probability they are a non-smoker?

Objective 2 – Explore the First Law of Probability

<u>Key Terms</u>

1st Law of Probability	
Independent events	
Dependent events	

<u>Summary</u>

We are now ready to explore the 1st law of probability which is nothing more than a rearrangement of the formula for computing **conditional probabilities**.

The **1st Law of Probability** says that the probability of event A and event B both occurring is:

$$\mathbf{P}(A \textbf{ AND } B) = \mathbf{P}(A) \cdot \mathbf{P}(B|A)$$

If A and B are *independent* events then $\mathbf{P}(B|A) = \mathbf{P}(B)$ and the formula is:

$$\mathbf{P}(A \textbf{ AND } B) = \mathbf{P}(A) \cdot \mathbf{P}(B)$$

The key is if you want the probability of this **and** that, multiply! The difference between dependent and independent events requires some care, and is best illustrated with an example.

Consider the following two situations:

1. You are having a party and believe the probability of your friend Rav attending, $P(Rav) = \dfrac{1}{5}$, and the probability of your friend Suk attending, $P(Suk) = \dfrac{2}{3}$. Rav and Suk do not know each other. What is the probability they both attend?

2. The probability of a sale at LL Bean, $P(Sale) = 20\%$, and the probability you spend more than \$100 at LL Bean, $P(Spend > \$100) = 15\%$. You are much more likely to spend over \$100 given there is a sale, $P(Spend > \$100 \mid Sale) = 60\%$. What is the probability you spend over \$100 and there is a sale?

Try to determine which situation involves independent events and which dependent events.

1. Given they do not know each other these are independent events and we can use the 1st law of probability for independent events:

$$P(Rav \text{ AND } Suk) = P(Rav) \cdot P(Suk) = \frac{1}{5} \cdot \frac{2}{3} = \frac{2}{15}$$

2. The sale and you spending over \$100 are not independent, so we use the First Law of Probability for dependent events:

$$P(Sale \text{ AND } Spend > \$100) =$$

$$P(Sale) \cdot P(Spend > \$100 \mid Sale) = 20\% \cdot 60\% = 12\%.$$

Note that for dependent events, we needed to choose the correct conditional probability.

As you read, jot down notes and questions. At the end of this section's Guided Worksheets there is a space for **Reflection** *and* **Monitoring Your Understanding***, where you can try and answer your questions after completing the guided activities and homework.*

Notes & Questions

A sub-committee of 2 people will be chosen from a group of 6: {Atia, Nasteho, Journey, Tuan, Espoir, Ryan}. The first 3 listed are women and the last 3 listed are men.

1. What is the sample space for this problem?

	☐	☐	☐	☐	☐	☐
☐		☐	☐	☐	☐	☐
☐			☐	☐	☐	☐
☐				☐	☐	☐
☐					☐	☐
☐						☐
☐						

2. Compute the **probability** the sub-committee contains 2 women using the 1st law.

3. You invite Nasteho and Tuan to attend an event. The probability of Nasteho attending is 20% while the probability of Tuan attending is 15%, but given Nasteho is attending the probability of Tuan attending increases to 50%. What is the probability that both Nasteho and Tuan attend?

4. You meet someone at the coffee shop who is reading a book on zen poetry, is vegan, and owns a rabbit. What is more likely: she is a banker, or she is a banker who does yoga?

Excel Preview – 11.6 Rain

Most weather websites report a probability of precipitation, but what exactly does it mean when you see a 40% chance of rain in your town? This probability is computed by multiplying the chance of rain somewhere in your town (confidence) and the percentage of area they estimate in your town that will get rained on (area). For example, if we are 50% confident of rain somewhere in your town AND we estimate 80% of your town will get wet given it actually does rain, THEN there is a 50%*80% = 40% chance that it will rain AND you will get wet!

1. In the description above write out the conditional probability using proper notation that you will get wet given it rains.

2. Are the events that it rains and that you get wet independent or dependent? Why?

3. How is the first law of probability used in computing the chance of rain tomorrow?

4. Do you think using the first law to compute the chance of rain tomorrow and you getting wet is better than just saying there is a 50% chance of rain somewhere in town tomorrow? Explain!

Section 11.3 Reflections
Monitor Your Understanding

Can you explain what a conditional probability is?

Objective 1 – Explore the Second Law of Probability
Objective 2 – Explore the Third Law of Probability

Objective 1 – Explore the Second Law of Probability

Key Terms

Union	
2nd Law of Probability	

Summary

In chapter 10 we saw that Excel has built-in logic functions: AND, OR, and NOT. The **1st Law of Probability** deals with the AND situation, the **probability** of this **event** *and* that event occurring (multiply!). · Next we will explore the probability of this event *or* that event occurring (add!).

Given the **Venn diagram** for the skin cream experiment in **Figure 11.8**, we can compute the probability of a participant having used the cream **OR** gotten better.

FIGURE 11.8 Venn Diagram for Skin Cream Experiment

Note that there are 223 people in the intersection of the two subsets. These are people who used the cream AND got better.

This event consists of the **union** of the two **subsets**, *Used Cream* $\cup$ *Got Better*. There are a total of $75+223+107=405$ people in the union, out of 426 total participants. Thus

$$\mathbf{P}\left(\text{Used Cream} \cup \text{Got Better}\right) = \frac{405}{426} = 95.1\%.$$

CAUTION! If we add the elements of both subsets together we double count the intersection. There are 298 people who used the cream and 330 who got better, adding these would give 628 people. We need to subtract the intersection to get the correct number in the union: $298+330-223=405$.

CAUTION! We often say things like, "I will go for a run or play golf." implying one or the other but not both. In mathematics *or* always means "this or that or both."

If "both" is impossible (for example rolling two dice and getting a 7 and an 11; or being a smoker and a non-smoker), we say *A* and *B* are disjoint and the probability of *A* and *B* (the intersection) is zero. In these situations we can just add the probabilities in the 2nd law (this will work for more than two events). For more than two events with non-empty **intersections**, the intersections that have you to subtract from the sum of individual probabilities gets complicated, and we will instead always use the third law.

*As you read, jot down notes and questions. At the end of this section's Guided Worksheets there is a space for **Reflection** and **Monitoring Your Understanding**, where you can try and answer your questions after completing the guided activities and homework.*

Notes & Questions

Guided Practice Activity #1 – 2nd Law

A sub-committee of 2 people will be chosen from a group of 6: { Atia, Nasteho, Journey, Tuan, Espoir, Ryan}. The first 3 listed are women and the last 3 listed are men.

1. What is the sample space for this problem?

	☐	☐	☐	☐	☐	☐
☐		☐	☐	☐	☐	☐
☐			☐	☐	☐	☐
☐				☐	☐	☐
☐					☐	☐
☐						☐
☐						

2. What is the **probability** that at least one man is chosen?

3. Compute the **probability** the sub-committee contains 1 man and 1 woman using the 1st law and the 2nd law.

4. You go to *Big E's Burrito's* in Brunswick, ME where 20% of the time you add sweet potato, 40% of the time you get corn salsa, and 5% of the time you get both.

   a. Are getting sweet potato and getting corn salsa disjoint events? Explain!

   b. What is the probability you get sweet potato or corn salsa?

5. Use the following probability distribution for rolling two dice to answer the questions.

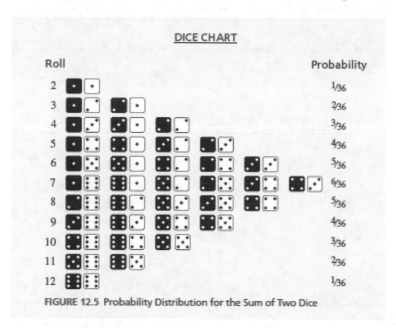

FIGURE 12.5 Probability Distribution for the Sum of Two Dice

a. What is the probability of rolling a sum of 7 and having at least one 2?

b. What is the sample space of having at least one 2?

c. What is the probability of rolling a sum of 7 or having at least one 2?

d. Which of the following pairs of events are disjoint?
   (sum of 10, at least one 3); (sum of 10, at least one 5); (sum of 10, sum of 5);
      (1st die is 2, sum is 5)

e. Find the probabilities of the events:
   P(sum of 10 OR at least one 3); **P**(sum of 10 OR at least one 5);
   P(sum of 10 OR sum of 5); **P**(1st die is 2 OR sum is 5)

Objective 2 – Explore the Third Law of Probability

Key Terms

Complement	
3rd Law of Probability	
Golden Rule of Probability	

Summary

There is a 3rd way to compute the **probability** of choosing at least 1 man for a sub-committee of 2 people chosen from a group of 6: {Atia, Nasteho, Journey, Tuan, Espoir, Ryan}. Using the fact that either an **event** occurs or it does not occur but not both: $P(A) + P(\text{NOT } A) = 1$. Thus, we can instead compute the probability of NOT getting at least 1 man, which is the same as the probability of getting two women: $P(At\ Least\ 1M) + P(2W) = 1$. Looking at **Table 11.7** we have $P(2W) = 3:15 = 20\%$, so $P(At\ Least\ 1M) = 80\%$.

The complement of an event A is the event that A does not occur, Not A. An event and its complement make up the entire sample space, so the sum of their probabilities must be 1.

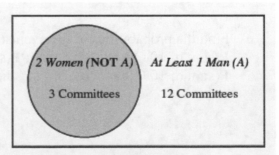

FIGURE 11.10 Venn Diagram for Committees of Two People

The complement of a union is the intersection of the complements. For example, the complement of the event, *Used Cream* OR *Got Better*, consists of participants who did not use the cream AND did not get better:

$$\mathbf{P}\big(\text{NOT}(\textit{Used Cream } \mathbf{OR} \textit{ Got Better})\big) = \mathbf{P}\big(\text{NOT}(\textit{Used Cream}) \textbf{ AND } \text{NOT }(\textit{Got Better})\big)$$

Modern probability theory has its roots in using the third law, and the Golden Rule in particular, to determine the correct probability for a dice game. In the excellent book, *Chance*, by Amir Aczel, the gambling paradox of Chevalier de Mere is recounted. The Chevalier enjoyed playing two different dice games in the casinos of 17th century Europe. The first game consisted of rolling 4 dice and winning if at least one "ace" (the number one) showed up. The Chevalier incorrectly reasoned that in the first game

$$\mathbf{P}\big(\textit{1st Ace or 2nd Ace or 3rd Ace or 4th Ace}\big) = \mathbf{P}\big(\textit{1st Ace}\big) + \mathbf{P}\big(\textit{2nd Ace}\big)$$

$$+ \mathbf{P}\big(\textit{3rd Ace}\big) + \mathbf{P}\big(\textit{4th Ace}\big) = 4:6 \text{ or } 67\%.$$

In this game the Chevalier was making the common mistake of forgetting to subtract the **probability** of the **intersections** when using the **2ⁿᵈ law**. His flawed logic would have us believe that if you rolled six dice you were guaranteed to get an ace since 6 times one-sixth = 100%. The second law is difficult to apply here since the sample space has $6^4 = 1{,}296$ possible outcomes and the intersections are too complicated to easily determine. Instead the **3ʳᵈ law** and in particular the **Golden Rule** (each die is independent of the others) readily provide the solution:

$$\mathbf{P}\big(\textit{At least one Ace}\big) = 1 - \mathbf{P}\big(\textit{No Aces}\big)$$

$$\mathbf{P}\big(\textit{At least one Ace}\big) = 1 - \mathbf{P}\big(\textit{No Ace } \mathbf{AND} \textit{ No ace } \mathbf{AND} \textit{ No ace } \mathbf{AND} \textit{ No ace}\big)$$

$$= 1 - \frac{5}{6} \times \frac{5}{6} \times \frac{5}{6} \times \frac{5}{6} = 1 - \left(\frac{5}{6}\right)^4 = 51.8\%$$

As you read, jot down notes and questions. At the end of this section's Guided Worksheets there is a space for **Reflection** *and* **Monitoring Your Understanding**, *where you can try and answer your questions after completing the guided activities and homework.*

Notes & Questions

Use the appropriate laws to solve the following:

1. Flip two coins and determine the probability of getting two heads (*2H*) and the probability of getting at least one head (*At least 1H*).

2. You are applying to 12 colleges and believe you have a small 8% **chance** of getting into any one of them. What is the **probability** of getting into at least one of the schools?

3. You go to *Big E's Burrito's* in Brunswick, ME where 20% of the time you add sweet potato, 40% of the time you get corn salsa, and 5% of the time you get both. What is the probability that you don't get sweet potato or corn salsa?

4. You roll a pair of dice four times. What is the probability you get at least one sum of 7?

In the chapter we saw an example recounting two dice games from the 17th century. The first game consisted of rolling 4 dice and winning if you get at least one "ace" (the number one).

	A	B	C	D	E	F	G	H	I	J
1		a.)	a.)	a.)	a.)	b.)				
2	1	Roll #1	Roll #2	Roll #3	Roll #4	# Aces		# 0 Aces	481	c.)
3	2	3	4	5	5	0		# At Least 1 Ace	519	c.)
4	3	1	2	3	4	1		P(At Least 1 Ace)	51.9%	c.)
5	4	1	3	5	4	1				
6	5	5	5	2	4	0				
7	6	6	5	1	1	2				

1. What function can be entered in cell **B3** to roll a fair six sided die?

2. What function can be entered in cell **F3** to count the number of 1's?

3. Assuming we have created 1000 simulations of the game what formula is in cell **I2** to count the number of times we lose (get 0 Aces)?

4. What formula is in cell **I3** to count the number of times we win (get at least 1 Ace)?

5. What formula is in cell **I4** to compute the empirical probability of winning in 1000 games?

6. Recall that the empirical probability will only approximate the true theoretical probability of 51.8%. The empirical probability will be above and below this true value, hovering like moths about a guiding light. How high do you think the empirical probability can get? 55%? 60%? 70%? Higher?

Monitor Your Understanding

Can you compare and contrast all three laws of probability?

| **Objective 1** – Understand Bayes' Theorem |
| **Objective 2** – Explore False Positives |

| **Objective 1** – Understand Bayes' Theorem |

Key Terms

Bayes' Formula	
Prior	
Posterior	

Summary

In the late 1740's the Reverend Thomas Bayes discovered the formula involving **conditional probabilities** that would subsequently be named after him. At the time he was interested in cause and effect, and wished to determine a way to quantify the **probability** of a cause given an observed effect. For example, we know that smoking is a possible cause of heart attacks, and in **Table 11.1** looked at some probabilities of a heart attack death given smoking, $\mathbf{P}\left(HA|Smoker\right)$.

Bayes was interested in the inverse probability, given a heart attack death, what is the probability of that person being a smoker, $\mathbf{P}\left(HA|Smoker\right)$? Clearly heart attacks don't cause smoking, so this is like working backwards from known data (heart attack) to likely causes (smoking). Bayes in particular wished to work backwards from observing the effect of creation to the probability of its likely cause being a creator. His simple theorem would come to divide generations of statisticians, with Bayesian adherents becoming bitter rivals of the frequentist camp. An excellent historical overview of this and the rise to ascendency of Bayes' Theorem in the modern era (due to increased computational power) is given in Sharon McGrayne's fabulous book, *The Theory That Would Not Die*.

We will start with a familiar example to derive the formula and then move on to more sophisticated applications. Recall from earlier that we computed the two **probabilities** $\mathbf{P}\left(Got\ Better|Used\ Cream\right)=74.8\%$ and $\mathbf{P}\left(Used\ Cream|Got\ Better\right)=67.6\%$ by looking at the information given in **Table 11.9** and **Figure 11.12**.

TABLE 11.9 Skin Cream Two-Way Table

	Rash Got Better	Rash Got Worse	Totals
Did use cream	223	75	298
Did not use cream	107	21	128
Totals	330	96	426

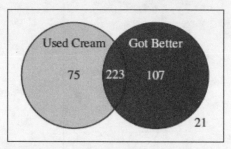

FIGURE 11.12 Venn Diagram for Skin Cream Experiment

Clearly using the cream would be the "cause" of the effect, getting better. We are interested in working backwards from effect to cause and will derive a new formula for computing $\mathbf{P}\left(Used\ Cream|Got\ Better\right)=\mathbf{P}\left(UC|GB\right)$.

In general the effect will be referred to as the data (D) and the cause will be referred to as the hypothesis (H). We wish to compute $\mathbf{P}\left(H|D\right)$. **Table 11.10** lists the two competing hypotheses given the data that someone got better: either they used the cream or did not.

> We know that 70% of the participants used the cream. Given the data that someone got better reduces this prior probability to 67.6%.

TABLE 11.10 Skin Cream Bayes' Table Given the Data That Someone Got Better

Hypotheses (H_i)	Priors $P(H_i)$	Likelihoods $P(D \mid H_i)$	Products $P(H_i) \cdot P(D \mid H_i)$	Posteriors $P(H_i \mid D)$
Did use cream	298/426 = 70.0%	223/298	223/426	223/330 = 67.6%
Did not use cream	128/426 = 30.0%	107/128	107/426	107/330 = 32.4%
Totals	100%		330/426	100%

The **priors** are the probabilities that someone did (70%) or did not (30%) use cream, and we update these probabilities to the **posteriors** given the data that someone got better. The **likelihoods** are the **conditional probabilities** $\mathbf{P}\left(D|H\right)$ which are multiplied by the priors in the numerator and denominator of the formula. The denominator in the formula is the sum of the products:

$$P(H_i|D) = \frac{P(H_i) \cdot P(D|H_i)}{\sum_i P(D|H_i) \cdot P(H_i)}$$

The table honestly makes things much easier for us. We can see that the posterior probability, 67.6%, of someone using the cream given they got better is less than the prior probability, 70%, of them using the cream. Clearly using the cream is not making it more likely they get better!

As you read, jot down notes and questions. At the end of this section's Guided Worksheets there is a space for **Reflection** *and* **Monitoring Your Understanding**, *where you can try and answer your questions after completing the guided activities and homework.*

Notes & Questions

Guided Practice Activity #1 – Skin Cream

In an experiment to determine if a skin cream helps a rash, the following data is collected:

	Rash got Better	Rash got Worse	Totals
Did use cream	223	75	**298**
Did not use cream	107	21	**128**
Totals	**330**	**96**	**426**

1. Does the skin cream help the rash? How do you know?

2. Draw a Venn Diagram of this information. Which representation is easier for you to process?

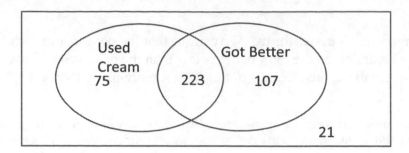

3. Compute and compare the probability of someone getting better given they used the cream, $\mathbf{P}(\textit{Got Better}|\textit{Used Cream}) = \mathbf{P}(GB|UC)$, and the probability of someone having used the cream given they got better, $\mathbf{P}(\textit{Used Cream}|\textit{Got Better}) = \mathbf{P}(UC|GB)$.

4. Which representation of the information, two-way table or Venn diagram, was more helpful to you in computing the probabilities?

5. Do you know how these two probabilities are related?

Starting with the formula for conditional probability:

$$\mathbf{P}(UC|GB) = \frac{\mathbf{P}(UC \textbf{ AND } GB)}{\mathbf{P}(GB)} = \frac{\mathbf{P}(UC) \cdot \mathbf{P}(GB|UC)}{\mathbf{P}(GB)} = \frac{\mathbf{P}(UC)}{\mathbf{P}(GB)} \cdot \mathbf{P}(GB|UC)$$

Now the people who got better (*GB*) either used cream (*UC*) or did not (*NoC*) so:

$$\mathbf{P}(GB) = \mathbf{P}(GB \cap UC) + \mathbf{P}(GB \cap NoC)$$

$$\mathbf{P}(GB) = \mathbf{P}(GB|UC) \cdot \mathbf{P}(UC) + \mathbf{P}(GB|NoC) \cdot \mathbf{P}(NoC)$$

Substituting in for the denominator above gives us Bayes' Formula:

$$P(UC|GB) = \frac{P(UC) \cdot P(GB|UC)}{P(GB|UC) \cdot P(UC) + P(GB|NoC) \cdot P(NoC)}$$

6. Recompute $P(UC|GB)$ using Bayes' Formula above.

Now see if you can follow how the following Bayes Table was constructed given the data that someone got better:

Hypotheses (H_i)	Priors $P(H_i)$	Likelihoods $P(D|H_i)$	Products $P(H_i) \cdot P(D|H_i)$	Posteriors $P(H_i|D)$
Did use cream	$\frac{298}{426} = 70.0\%$	$\frac{223}{298}$	$\frac{223}{426}$	$\frac{223}{330} = 67.6\%$
Did not use cream	$\frac{128}{426} = 30.0\%$	$\frac{107}{128}$	$\frac{107}{426}$	$\frac{107}{330} = 32.4\%$
Totals	100%		$\frac{330}{426}$	100%

How does the table indicate whether or not the cream works?

Guided Practice Activity #2 – Smoking Apples to the Full Monty

Your father is 55 years old and has just had a heart attack. He says he is not smoking but you think there is a 20% chance he has been smoking. Given that he has had a heart attack what is the new probability that he is smoking? Our two hypotheses are that he is smoking (S) or not smoking (NS) and Table 11.1 from the text gives us the likelihood of a heart attack (A) in both cases: $P(A|S) = 4.1\%$ and $P(A|NS) = 1.9\%$.

1. Fill in a Bayes Table and compute the new probability of smoking given the new data of a heart attack:

Hypotheses (H_i)	Priors $P(H_i)$	Likelihoods $P(D\|H_i)$	Products $P(H_i) \cdot P(D\|H_i)$	Posteriors $P(H_i\|D)$
Smoking	[ ]	[ ]	[ ]	[ ]
Not Smoking	[ ]	[ ]	[ ]	[ ]
Totals	100%		[ ]	100%

Apply Bayes' Formula with 3 Hypotheses

Three siblings (Galen, Mei, and Beau) are equally likely to eat an apple, but the probability they

leave the core on the counter differs: $\mathbf{P}(Core|Galen) = \dfrac{1}{5}$, $\mathbf{P}(Core|Mei) = \dfrac{3}{5}$, and

$\mathbf{P}(Core|Beau) = \dfrac{4}{5}$. If we find a core on the counter, what is the new probability that each

sibling was the culprit? Our three hypotheses are that each sibling ate the apple and left the core on the counter.

2. Fill in a Bayes Table and compute the new probabilities of each sibling eating the apple given the new data of a core:

Hypotheses (H_i)	Priors $P(H_i)$	Likelihoods $P(D\|H_i)$	Products $P(H_i) \cdot P(D\|H_i)$	Posteriors $P(H_i\|D)$
Galen	[ ]	[ ]	[ ]	[ ]
Mei	[ ]	[ ]	[ ]	[ ]
Beau	[ ]	[ ]	[ ]	[ ]
Totals	100%		[ ]	100%

Solve the Monty Hall Problem

The Monty Hall Problem is a classic probability conundrum. It has fooled experts with advanced degrees in statistics and led to public embarrassment as these experts attacked each other in print. The problem states that you are on a game show and must choose 1 out of 3 doors. There is a car behind one of the doors. After you have selected your door, the host, Monty Hall, opens another door that does not have the car (he knows where the car is) and asks if you want to change your door. Should you switch? Our three hypotheses are that the car is equally likely to be behind one of the 3 doors. Let's assume you pick door A and Monty opens door B, should you switch to door C?

3. Compute the likelihood that Monty opens door B given the data that the car is behind door A.

4. Compute the likelihood that Monty opens door B given the data that the car is behind door B.

5. Compute the likelihood that Monty opens door B given the data that the car is behind door C and fill in the table.

Hypotheses (H_i)	Priors $P(H_i)$	Likelihoods $P(D\|H_i)$	Products $P(H_i) \cdot P(D\|H_i)$	Posteriors $P(H_i\|D)$
Door A				
Door B				
Door C				
Totals	100%			100%

Key Terms

False positive	
False negative	

Summary

One last application of **Bayes' Formula** to making informed health decisions. There exist many tests for diseases. A **false positive** indicates you test positive for the disease (indicating sick) but don't actually have it, while a **false negative** indicates you test negative for the disease (indicating healthy) but actually have it.

As you read, jot down notes and questions. At the end of this section's Guided Worksheets there is a space for **Reflection** *and* **Monitoring Your Understanding***, where you can try and answer your questions after completing the guided activities and homework.*

Notes & Questions

Guided Practice Activity #1 – False Positives

A certain disease occurs in 5 out of 1,000 people. A test has a false positive rate of 3% and a false negative rate of 1%. What is the probability that a person who tests positive actually has the disease, $\mathbf{P}\left(Sick\middle|+\right)$?

1. What is the probability that a person actually has the disease, $\mathbf{P}(Sick)$?

2. What is the probability that a person does not have the disease, $P(Well)$?

3. What is the probability that a person tests positive given they actually have the disease, $P(+|Sick)$?

4. What is the probability that a person tests positive given they actually have the disease, $P(+|Well)$?

5. Now fill in the Bayes table to determine the probability that a person who tests positive actually has the disease, $P(Sick|+)$:

Hypotheses (H_i)	Priors $P(H_i)$	Likelihoods $P(D|H_i)$	Products $P(H_i) \cdot P(D|H_i)$	Posteriors $P(H_i|D)$
Sick				
Well				
Totals	100%			100%

Excel Preview – 11.8 Bayes

The table gives educational attainment data by race and ethnicity. We will use this data to practice Bayesian analysis. **Given someone has received a Bachelor's degree, what is the probability they are White/Black/or Asian?** We will assume the prior probabilities are just the proportion of each race (White, Black, and Asian) in the population; and fill in the table to update these priors to new posterior probabilities that take into account the data.

Table 3. Detailed Years of School Completed by People 25 Years and Over by Sex, Age Groups, Race and Hispanic Origin: 2015					
(Numbers in thousands. Civilian noninstitutionalized population /Excluding members of the Armed Forces living in barracks.)					
Source: U.S. Census Bureau, Current Population Survey, 2015 Annual Social and Economic Supplement					
	White	Non-Hispanic White	Black	Asian	Hispanic (any race)
	168,420	140,638	25,420	12,331	31,020
High School (Diploma)	44,708	37,224	7,870	2,247	8,277
College (no Degree)	27,558	23,743	5,200	1,183	4,397
Associate's Degree (Vocational)	7,474	6,595	1,123	302	983
Associate's Degree (Academic)	9,467	8,369	1,409	492	1,242
Bachelors degree	34,917	31,927	3,636	4,013	3,338
Master's degree	14,706	13,776	1,669	1,776	1,085
Professional degree	2,795	2,595	197	330	209
Doctorate degree	2,803	2,652	211	532	169

1. First compute the total number of Whites, Blacks and Asians in the population. Note that Hispanic is an ethnicity so Hispanics can be any of these three races.

2. Second compute the prior probabilities which are just going to be the proportion of each race in the population. Fill in table and format as percentage to 1 decimal place.

Data = Bachelor's Degree	b.)	c.)	d.)	e.)
Hypotheses (H_i)	Priors	Likelihoods	Products	Posteriors
	$P(H_i)$	$P(D\|H_i)$	$P(H_i) \times P(D\|H_i)$	$P(H_i\|D)$
White	81.7%	20.7%	16.9%	82.0%
Black	12.3%	14.3%	1.8%	8.5%
Asian	6.0%	32.5%	1.9%	9.4%
Totals	100%		20.6%	100%

3. Third compute the likelihoods which are conditional probabilities of the data given each hypothesis respectively.

4. Compute the products and sum them.

5. Compute the posteriors which are conditional probabilities of each hypothesis respectively given the data using Bayes' Formula which simply divides each product by the sum.

Section 11.5 Reflections
Monitor Your Understanding

Can you explain the power and applications of Bayes' Theorem?

Chapter 12 – Statistically Speaking

Section 12.1 Data Distributions: Frequency and Probability

Objective 1 – Explore Frequency Distributions
Objective 2 – Explore Probability Distributions

Objective 1 – Explore Frequency Distributions

Key Terms

Data	
Quantitative data	
Qualitative data	
Discrete data	
Continuous data	
Essences	
Shape of data	
Center of data	
Spread of data	
Frequency distribution	

Sample	
Population	
Sample statistics	
Population parameters	

Summary

The discipline of **statistics** can be described as the study of variability in **data**. If everyone was the same height and every student got the same SAT score we would not have the science of statistics. You ask people leaving the dining hall how much cash they have on hand and collect the following data. The data is difficult to analyze because it is not ordered.

$4	$8	$0	$12	$43	$3	$2	$12	$27
$35	$13	$18	$20	$20	$10	$14	$35	$5
$40	$7	$10	$22	$15	$10	$5	$30	$25
$20	$10	$15	$12					

Data: Cash on Hand

We can create the frequency distribution by using **bins** of length $10: 0-9, 10-19, etc. and count how many values are in each bin. The **histogram** creates columns with a height equal to these counts or frequencies.

Values	Frequency
$0…$9	8
$10…$19	12
$20…$29	6
$30…$39	3
$40…$49	2

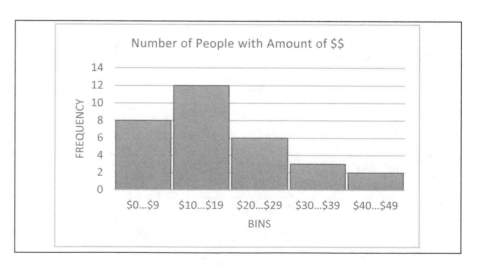

The **frequency distribution** obscures individual data values but gives an overall sense of the **shape**, **center** and **spread** of the **data** (the **essences**). Using Excel's built-in functions we can readily compute the **descriptive statistics:**

MEAN	MEDIAN	MODE	STDEV.S	MAX	MIN	RANGE	COUNT
16.2	13	10	11.4	43	0	43	31

The average amount of cash is $16.20, with half the people having more than $13, and the most common amount is $10. The average distance from the mean is $11.40, with a maximum value of $43 and a minimum of $0. We are using **STDEV.S** because we have sample data.

The previous example illustrated **sample data**, 31 people, out of the entire **population** of people on campus. It usually is not feasible to collect data for an entire population, so we use **sample statistics** to estimate the associated **population parameters**.

	Count	Mean	Standard Deviation
Sample Statistics	n	$\overline{X} = \dfrac{\sum x_i}{n}$	$s = \sqrt{\dfrac{\sum\left(x_i - \overline{X}\right)^2}{n-1}}$
Populations Parameters	N	$\mu = \dfrac{\sum x_i}{N}$	$\sigma = \sqrt{\dfrac{\sum\left(x_i - \mu\right)^2}{N}}$

As you read, jot down notes and questions. At the end of this section's Guided Worksheets there is a space for **Reflection** *and* **Monitoring Your Understanding***, where you can try and answer your questions after completing the guided activities and homework.*

Notes & Questions

Guided Practice Activity #1 – Number of Cars

You sample 30 people on the quad and ask them how many cars they have at home.

Number of Cars Data					
2	3	0	1	1	1
0	2	3	1	2	1
4	1	1	2	1	2
1	1	0	2	3	2
1	0	1	4	5	2

1. Fill in the Frequency Table to get the distribution.

Number of Cars	Frequency
0	☐
1	☐
2	☐
3	☐
4	☐
5	☐

2. Create a histogram for the frequencies.

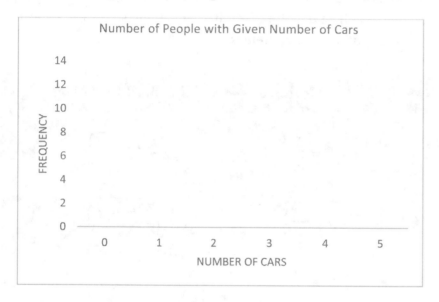

3. Compute the descriptive statistics:

a. Measures of Center:
   - Mode:
   - Median:
   - Mean:

b. Measures of Spread:
   - Max:
   - Min:
   - Range:
   - Count:

c. The standard deviation is 1.24. Use this sample statistic in a sentence with the mean.

Objective 2 – Explore Probability Distributions

Key Terms

Probability distribution	
Expected value	
Relative frequency	

Summary

We are now going to explore distributions for probabilistic events. We often ask questions like: "What is the chance a city has a cost of living index less than 90?" or "What percentage of people carry more than $30 cash?" This requires interpreting our **distributions** not as counts but as **relative frequencies**, or percentages of data in a given bin.

People are asked to rate a product on a scale of 1 – 5. Given the data shown, we can see that 30 people participated and 6 out of 30 gave a rating less than 3, or 20.0%.

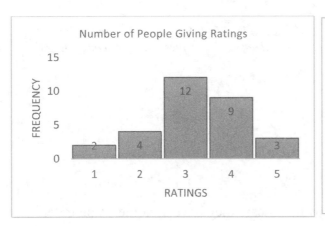
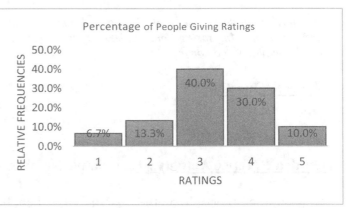

We can see from the **frequency distribution** that two people gave a rating of 1, four a rating of 2, twelve a rating of 3 etc. Alternatively we can see from the **probability distribution** that 6.7% of the people gave a rating of 1, 13.3% a 2, 40% a 3, etc.

Rating	Frequency	Relative frequency
1	2	6.7%
2	4	13.3%
3	12	40.0%
4	9	30.0%
5	3	10.0%

To compute the **mean** we can add up all the ratings and divide by 30, which is equivalent to:

$$\frac{1+1+2+\cdots+3+\cdots+4+\cdots+5}{30} = \frac{1\cdot2+2\cdot4+3\cdot12+4\cdot9+5\cdot3}{30}$$

$$= 1\cdot\frac{2}{30}+2\cdot\frac{4}{30}+3\cdot\frac{12}{30}+4\cdot\frac{9}{30}+5\cdot\frac{3}{30}$$

The **mean** for a probability distribution is given by:

$$\mu = \frac{\sum x_i}{N} = \frac{\sum x_k \cdot f_k}{N} = \sum x_k \cdot P_k = 3.23$$

CAUTION! We switch subscripts from i to k because the first sums over all 30 data values, while the second sums over just the 5 unique data values. Note how nice the formula is that uses the relative frequencies or probabilities! The **SUMPRODUCT** function in Excel allows us to sum the products in one formula. Also the count disappears, which is why we use **population** formulas. There exists a similar formula for the standard deviation for a probability distribution:

$$\sigma = \sqrt{\frac{\sum(x_i-\mu)^2}{N}} = \sqrt{\frac{\sum(x_k-\mu)^2\cdot f_k}{N}} = \sqrt{\sum(x_k-\mu)^2\cdot P_k} = 1.02$$

The average rating is 3.23 with an average distance away from the mean of 1.02.

As you read, jot down notes and questions. At the end of this section's Guided Worksheets there is a space for **Reflection** *and* **Monitoring Your Understanding***, where you can try and answer your questions after completing the guided activities and homework.*

Notes & Questions

Guided Practice Activity #1 – Relative Number of Cars

You sample 30 people on the quad and ask them how many cars they have at home. You enter the data in the Frequency Table to get the distribution.

1. Compute the relative frequencies.

Number of Cars	Frequency	Relative Frequency
0	4	☐
1	12	☐
2	8	☐
3	3	☐
4	2	☐
5	1	☐

2. Create histograms for both the frequencies and relative frequencies.

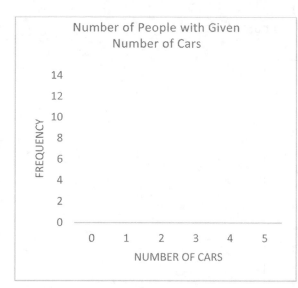

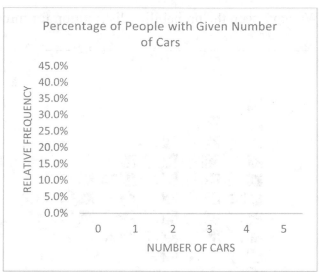

3. Compute the mean using the formula for a probability distribution:
    $$\sum x_k \cdot P_k =$$

4. Compute the standard deviation using the formula for a probability distribution:
    $$\sqrt{\sum (x_k - \mu)^2 \cdot P_k} =$$

Number of Cars	Frequency	$x_k - \mu$	$(x_k - \mu)^2$	Relative Frequency	$(x_k - \mu)^2 \cdot P_k$
0	4	[ ]	[ ]	[ ]	[ ]
1	12	[ ]	[ ]	[ ]	[ ]
2	8	[ ]	[ ]	[ ]	[ ]
3	3	[ ]	[ ]	[ ]	[ ]
4	2	[ ]	[ ]	[ ]	[ ]
5	1	[ ]	[ ]	[ ]	[ ]
				Sum:	[ ]
				SQRT:	[ ]

Guided Practice Activity #2 – Doubles

We have seen the probability distribution for rolling a pair of dice:

1. What is the probability of rolling doubles?

The **1st Law of Probability** says that the probability of event A and event B both occurring is:

$$\mathbf{P}(A \text{ AND } B) = \mathbf{P}(A) \cdot \mathbf{P}(B|A)$$

If A and B are *independent* events then $\mathbf{P}(B|A) = \mathbf{P}(B)$ giving: $\mathbf{P}(A \text{ AND } B) = \mathbf{P}(A) \cdot \mathbf{P}(B)$

2. What is the probability of rolling doubles on two consecutive rolls of a pair of dice?

The **3rd Law of Probability** says that the probability of event A is one minus the probability of its complement, **NOT** A:

$$\mathbf{P}(A) = 1 - \mathbf{P}(\text{NOT } A)$$

3. What is the probability of not rolling doubles?

4. What is the probability of rolling a pair of dice three times and getting doubles on the first roll, then not doubles on the next two rolls?

The **2nd Law of Probability** says that the probability of event A or event B occurring is:
$$P(A \text{ OR } B) = P(A) + P(B) - P(A \text{ AND } B)$$

If A and B are *disjoint* events then $P(A \text{ AND } B) = 0$ giving: $P(A \text{ OR } B) = P(A) + P(B)$

5. What is the probability of rolling a pair of dice three times and getting one doubles?

6. Next we are going to explore a game of chance that involves rolling a pair of dice 3 times. You will win a payout for rolling doubles, and lose money if you don't roll doubles according to the table shown below. First list the possible outcomes of this game and associated probabilities.

Roll 1	Roll 2	Roll 3	Outcome	Probability	# Doubles
		Doubles	☐	☐	☐
	Doubles				
Doubles $P = \dfrac{1}{6}$		X	☐	☐	☐
		Doubles	☐	☐	☐
	X				
		X	☐	☐	☐
		Doubles	☐	☐	☐
	Doubles				
X $P = \dfrac{5}{6}$		X	☐	☐	☐
		Doubles	☐	☐	☐
	X				
		X	☐	☐	☐

7. Fill in the probability table as % (1 decimal place) and a fraction:

Doubles	Probability	Fraction	Winnings
0	☐	☐	-$10.00
1	☐	☐	$5.00
2	☐	☐	$25.00
3	☐	☐	$250.00

8. Now compute the expected number of doubles in 3 rolls:

9. And the expected winnings in 3 rolls:

Excel Preview – 12.1 Pop_Density

Compute the descriptive statistics of the **population densities** for the 50 states and D.C. Use **STDEV.P** for the population standard deviation and format to one decimal place.

▲	A	B	C	D	E	F
1		Population (1,000's)	Population Density per square mile of land area			
2		2010	2010			
3	United States	308,745.5	87.4		a.)	
4					MEAN	MEDIAN	M(
5	Alabama	4,779.7	94.4		384.4	101.2
6	Alaska	710.2	1.2			
7	Arizona	6,392.0	56.3		b.)	
8	Arkansas	2,915.9	56.0			

1. What formula would be entered to compute the following statistics?
   a. Mean:
   b. Median:
   c. Mode:
   d. Standard deviation:
   e. Max:

    f. Min:

    g. Range:

    h. Count:

2. How can the mean be so much larger than the median?

3. To compute the Frequency table we can use the **COUNTIFS** function. The formula for the first frequency (Bin 0 … 25) would be =**COUNTIFS(C\$5:C\$55,">=0",C\$5:C\$55,"<25")**. What is the formula in the next cell?

Bins	Frequency	Midpoints	Rel. Freq.
0 … 25	9	12.5	
25 … 50	5	37.5	
50 … 75	7	62.5	
75 … 100	4	87.5	
100 … 250	15	175	
250 … 500	5	375	
500 … 10,000	6	4750	

4. Compute the relative frequencies in the table. Use them and the midpoints to estimate the mean.

$$\sum m_i \cdot P_i = 668.4$$

Section 12.1 Reflections
Monitor Your Understanding

Can you describe the difference between a frequency distribution and probability distribution?

| **Objective 1** – Understand Binomial Distributions |
| **Objective 2** – Understand Normal Distributions |

| **Objective 1** – Understand Binomial Distributions |

Key Terms

Binomial experiment	
Binomial distribution	
Binomial probability	

Summary

In section 12.1 in the e-text we explored rolling a pair of dice 3 times and counting the number of naturals (rolling 7 or 11). This is an example of a binomial experiment. Using $p = \dfrac{2}{9}$ for the probability of a success, we can compute the binomial probabilities for 3 trials. There are four possible outcomes: 0, 1, 2, and 3 successes in 3 trials. We use the **first law of probability** to compute the probabilities of specific outcomes. There is only one way to get three successes, but there are three ways to get two successes (the one failure could occur 1st, 2nd or 3rd).

$$P(S.S.S) = \frac{2}{9} \cdot \frac{2}{9} \cdot \frac{2}{9} = \frac{8}{79}$$

$$P(F.S.S) = P(S.F.S) = P(S.S.F) = \frac{2}{9} \cdot \frac{2}{9} \cdot \frac{7}{9} = \frac{28}{729}$$

Thus the **probability** of getting two naturals in 3 rolls of a pair of dice involves counting all the ways the two naturals can occur. There are three ways to choose two of the three spots to be naturals. Hopefully this language reminds you of **combinations**.

Now we can compute the probability of getting exactly two naturals:

$$P(2 \text{ Naturals}) = 3C2 \cdot \left(\frac{2}{9}\right)^2 \cdot \left(\frac{7}{9}\right)^1 = 3 \cdot \frac{28}{729} = \frac{84}{729} = 11.5\%$$

The other binomial probabilities are computed in a similar fashion:

$$P(0 \text{ Naturals}) = \mathbf{3C0} \cdot \left(\frac{2}{9}\right)^0 \cdot \left(\frac{7}{9}\right)^3 = \mathbf{1} \cdot \frac{343}{729} = \frac{343}{729} = 47.1\%$$

$$P(1 \text{ Naturals}) = \mathbf{3C1} \cdot \left(\frac{2}{9}\right)^1 \cdot \left(\frac{7}{9}\right)^2 = \mathbf{3} \cdot \frac{98}{729} = \frac{294}{729} = 40.3\%$$

$$P(2 \text{ Naturals}) = \mathbf{3C2} \cdot \left(\frac{2}{9}\right)^2 \cdot \left(\frac{7}{9}\right)^1 = \mathbf{3} \cdot \frac{28}{729} = \frac{84}{729} = 11.5\%$$

$$P(3 \text{ Naturals}) = \mathbf{3C3} \cdot \left(\frac{2}{9}\right)^3 \cdot \left(\frac{7}{9}\right)^0 = \mathbf{1} \cdot \frac{8}{729} = \frac{8}{729} = 1.1\%$$

We can create a **binomial distribution** with **histogram** for the number of naturals in 10 rolls of a pair of dice, and 100 rolls. We know the binomial probabilities for 10 rolls are of the form:

$$P(x \text{ Naturals}) = 10Cx \cdot \left(\frac{2}{9}\right)^x \cdot \left(\frac{7}{9}\right)^{10-x}$$

We will be using Excel so can enter this version of the formula:

$$P(x \text{ Naturals}) = \mathbf{COMBIN}(10, x) \cdot \left(\frac{2}{9}\right)^x \cdot \left(\frac{7}{9}\right)^{10-x}$$

We fill this formula down for the eleven possible outcomes for rolling a pair of dice ten times to create the **binomial distribution**:

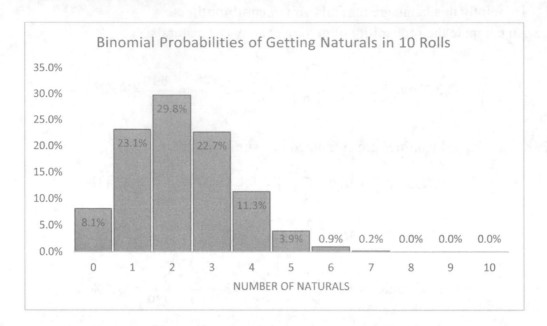

Now repeat the process for 100 rolls. Notice how symmetric and bell shaped the binomial distribution is becoming! As the number of rolls gets large this distribution approaches the normal distribution.

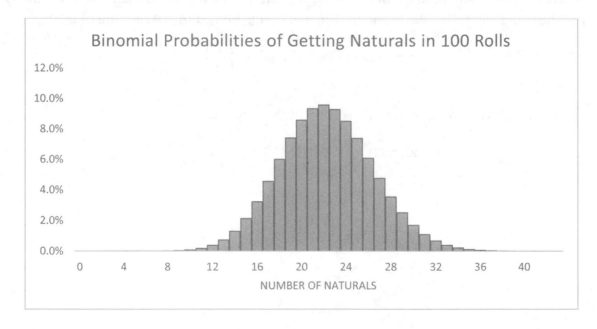

The mean is $\mu = n \cdot p = 22.2$ and $\sigma = \sqrt{n \cdot p \cdot q} = 4.16$ for 100 rolls.

*As you read, jot down notes and questions. At the end of this section's Guided Worksheets there is a space for **Reflection** and **Monitoring Your Understanding**, where you can try and answer your questions after completing the guided activities and homework.*

Notes & Questions

Guided Practice Activity #1 – Binomial Doubles

Rolling a pair of dice and hoping to get doubles can be considered a binomial experiment.

1. Assuming you roll the pair of dice 60 times what is the mean and standard deviation assuming getting doubles is a success?

2. What is the formula you can type into Excel using for the probability of getting x successes in n trials with two possible outcomes (success or failure).

3. Assuming you roll a pair of dice 60 times and getting doubles is a success, what is the probability of getting 15 doubles?

4. A data value is unusual if it lies more than two standard deviations from the mean, this is considered "far" from the mean. Is getting 15 doubles in this binomial experiment unusual?

5. What is smallest number of doubles you can get and not have it be unusual?

Key Terms

Normal distribution	
Standard normal distribution	
95% rule	
Probability density function (pdf)	
Unusual data value	

Summary

There are two main applications of **binomial distributions**:

1. **Normal Distributions**
2. Polling Questions

Normal distributions were introduced in the Spotlight on Statistics in Chapter 2.

CAUTION! The binomial distribution is approximately normal if $n \cdot p \geq 5$ and $n \cdot q \geq 5$.

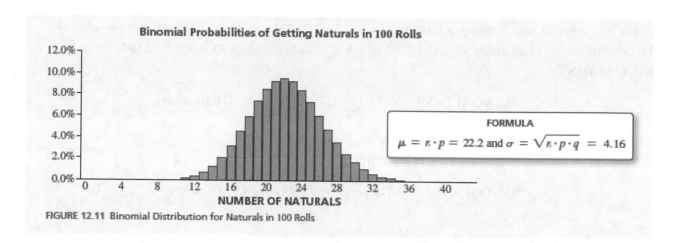

FIGURE 12.11 Binomial Distribution for Naturals in 100 Rolls

The familiar bell shaped normal curve is a theoretical model for the binomial distribution, as well as an approximation for many real world data sets (heights of people, housing prices, stock market price changes,…). Given the variety of possible units we replace our **data** values with **z-scores**, and talk about how many **standard deviations** away from the **mean** a value is.

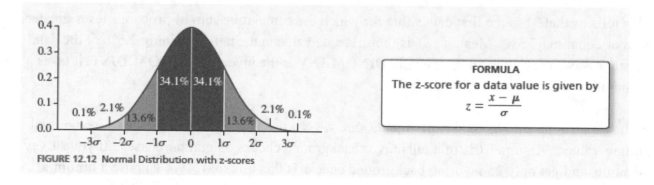

FIGURE 12.12 Normal Distribution with z-scores

CAUTION! The normal curve is the graph of a **probability density function** meaning the *y*-values represent the density or percentage of data per unit interval on the *x*-axis.

The fundamental question we will be asking is whether a given data value is "unusual". A data value is **unusual** if it lies more than two standard deviations from the mean, this is considered "far" from the mean. In a normal distribution this roughly means outside the 95% interval about the mean. Note that 2.5% of the data will lie in the left tail and 2.5% in the right tail.

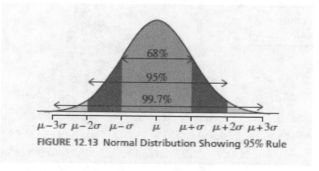

FIGURE 12.13 Normal Distribution Showing 95% Rule

Excel has built-in functions to determine the location of a data value in a binomial distribution. To determine the probability of rolling 30 or fewer naturals in 100 rolls we will use **BINOM.DIST**:

$$=\textbf{BINOM.DIST}\left(30,100,\frac{2}{9},0\right)=1.7\%=\textbf{P}(30 \text{ naturals})$$

We want the cumulative sum of all probabilities for 30 or less:

$$=\textbf{BINOM.DIST}\left(30,100,\frac{2}{9},1\right)=\textbf{97.4\%}=\textbf{P}(\leq30 \text{ naturals})$$

CAUTION! Note the difference between these two probabilities.
Another useful function is **BINOM.INV** which gives the data value corresponding to a cumulative **probability**:

$$=\textbf{BINOM.INV}\left(100,\frac{2}{9},0.975\right)=31$$

This tells us that 31 is the first data value for which the cumulative sum of probabilities is greater than or equal to 97.5%. Meaning 31 is the first data value in the tail containing 2.5% of the data! Another way to say 30 is **not** in the tail. **BINOM.INV** is the inverse to **BINOM.DIST,** it takes a probability and gives a data value.

Polling questions are binomial experiments, and we can use BINOM.DIST to determine unusual polling values. Assume 70% of adults favor background checks on gun purchases. If you survey 40 adults and get only 25 favoring background checks is this **unusual**? We can use a **binomial distribution** for this situation with $n = 40$ and $p = 0.7$.

$$=\textbf{BINOM.DIST}(25, 40, 0.7, 1) = \textbf{19.3\%}$$

Telling us that 19.3% of the data values are 25 or less, so clearly 25 is not far away in the tail containing only 2.5% of the data.

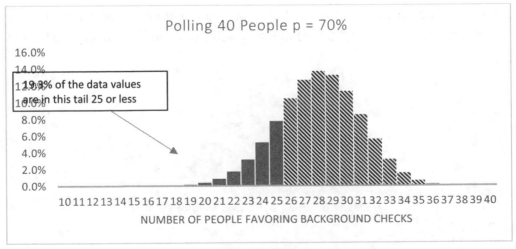

There exist similar functions for normal distributions with a given mean and standard deviation. **NORM.DIST** will give the cumulative percentage of data less than or equal to a given value and **NORM.INV** give the data value associated to a given cumulative percentage. There also are **NORM.S.DIST** and **NORM.S.INV** that work for the **standard normal distribution** with a mean of 0 and standard deviation of 1.

*As you read, jot down notes and questions. At the end of this section's Guided Worksheets there is a space for **Reflection** and **Monitoring Your Understanding**, where you can try and answer your questions after completing the guided activities and homework.*

Notes & Questions

Guided Practice Activity #1 – SAT Averages

Assuming the **distribution** of average SAT scores by institution is approximately **normal**, we have the descriptive statistics:

MEAN	MEDIAN	MODE	STDEV.P	MAX	MIN	RANGE	COUNT
1059.1	1039.5	1050	**133.4**	1545	720	825	1304

Use the functions **NORM.INV**(probability, mean, std_dev) and **NORM.DIST**(x, mean, std_dev, cumulative) to answer the following.

1. What function would we type into Excel to determine the cutoff SAT scores giving 2.5% of data in each tail?

2. What function would we type into Excel to determine if University of Michigan is **unusual** with an average SAT of 1380?

3. What does the output of the following functions tell us?

   a. =**NORM.DIST**(980, 1059.1, 133.4, 1) = 0.2766

   b. =**NORM.INV**(35%, 1059.1, 133.4) = 1007.69

<u>Excel Preview</u> – 12.5 Poll

This Excel problem uses the **distribution of sample proportions**, here we just use the binomial distribution for a poll. A June 2018 Gallup Poll of 1,520 U.S. adults found that a record high 75% think immigration is a good thing for the country.

1. Assuming the true proportion is 75% and a success is saying it is a good thing. If you survey 40 adults and ask the same question, what function will determine the probability that **25 or less** say they think immigration is a good thing? Use **BINOM.DIST**(number_s, trials, probability_s, cumulative).

2. Assuming the true proportion is 75%. If you survey 40 adults and ask the same question, what function will determine the probability that **exactly** 25 say they think immigration is a good thing?

3. Determine the mean and standard deviation of this binomial distribution surveying 40 people.

4. Can we assume this binomial distribution is approximately normal?

5. Determine the data values that would be unusual.

6. How could you use **BINOM.INV**(trials, probability_s, alpha) to determine these cutoff values?

<u>Section 12.2 Reflections</u>
Monitor Your Understanding

Can you describe how binomial distributions are related to normal distributions?

Section 12.3 Sampling Distributions and Confidence Intervals

Objective 1 – Understand Sampling Distributions
Objective 2 – Explore Confidence Intervals

Objective 1 – Understand Sampling Distributions

Key Terms

Sampling distribution	
Distribution of sample means	
Central Limit Theorem (CLT)	
Standard error	

Summary

In the SAT example we saw that University of Michigan had an **unusual** SAT average of 1380 because it was far away from the **mean** of 1059.1, a full 2.41 **standard deviations** ($\sigma = 133.4$) away with only 0.8% of schools having greater SAT averages. For comparison, Berea College at 1125 is only 0.49 standard deviations above the mean and thus not unusual. We now turn our attention to taking random **samples**. Often we do not have the capability of measuring every element in the population so we measure a random sample. The rest of this chapter will focus on making inferences from the sample statistics to the **population**. If we had sampled the average SAT scores of just 30 schools, how certain could we be that our sample mean is close to the unknown true population mean?

How likely is it for a random sample of 30 schools to have a sample mean of 1380? How about 1125?

This is a thought question, no computation. We have seen it is possible for a single school to have an average SAT score of 1380, but less than 1% of schools are out in this upper tail. So if we were randomly selecting schools we would have to select about half of our sample 30 schools from this upper tail to get the sample mean to be 1380. This seems highly unlikely, so unlikely that the **probability** is basically zero!

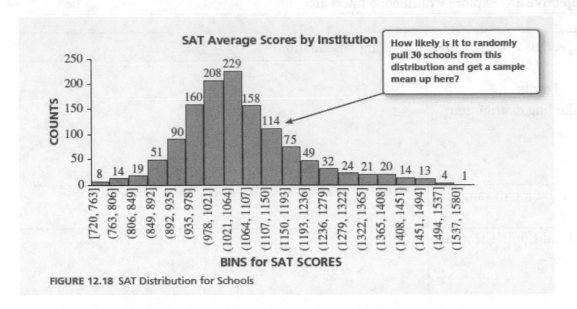

FIGURE 12.18 SAT Distribution for Schools

So there is zero **probability** of a random **sample** of 30 schools having a sample **mean** of 1380. We expect our samples to have some schools above the mean and some below the mean, basically averaging out and giving us a sample mean "pretty close" to the true **population** mean of 1059. In other words we expect our **sampling distribution** to be less spread out than the distribution of schools themselves.

So what about a random sample having a sample mean of 1125? This seems much more likely, but still would require us to have randomly pulled (selected) many schools in the right tail to get our sample mean so far away from 1059. To answer this question we need a result that tells us how spread out the distribution of sample means is compared to the original distribution. The **Central Limit Theorem** tells us that the **distribution of sample means** will have the same **mean** 1059.1 but will be less spread out:

$$\mu_{\bar{X}} = 1059.1 \quad \sigma_{\bar{X}} = \frac{133.4}{\sqrt{30}} = 24.3$$

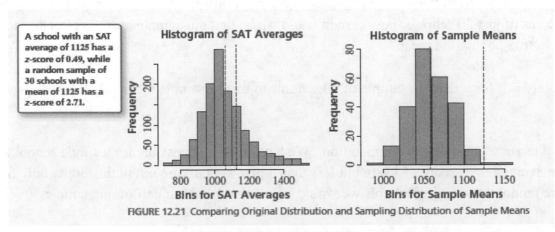

FIGURE 12.21 Comparing Original Distribution and Sampling Distribution of Sample Means

The **histograms** of both the original data of SAT averages for all 1304 institutions and the sampling distribution of 250 random samples of size 30 are shown on this page. The mean is the same for both distributions and shown in red dashed line. The value of 1125 is shown as the vertical blue line.

We can now determine that the z-score of a sample with a mean of 1125:

$$z = \frac{1125 - \mu_{\bar{X}}}{\sigma_{\bar{X}}} = \frac{1125 - 1059.1}{24.3} = 2.71$$

A random sample of 30 schools with a mean of 1125 has a z-score of 2.71. Thus, this is **unusual**!

As you read, jot down notes and questions. At the end of this section's Guided Worksheets there is a space for **Reflection** *and* **Monitoring Your Understanding***, where you can try and answer your questions after completing the guided activities and homework.*

Notes & Questions

Guided Practice Activity #1 – SAT Sample Averages

The **distribution** of average SAT scores by institution has the descriptive statistics:

MEAN	MEDIAN	MODE	STDEV.P	MAX	MIN	RANGE	COUNT
1059.1	1039.5	1050	**133.4**	1545	720	825	1304

Use the functions **NORM.INV**(probability, mean, std_dev) and **NORM.DIST**(x, mean, std_dev, cumulative) to answer the following where appropriate.

1. Assume we take samples of size 50. How do we know the distribution of sample means will be normal?

2. What is the mean and standard deviation of the distribution of sample means?

3. What is the *z*-score of a sample mean of 1100? Is this unusual?

4. Is it possible for a sample mean to be 960?

5. What functions would we type into Excel to find the cutoffs for unusual sample means?

6. What function would we type into Excel to determine the percentage of sample means less than 1000?

7. What does the output of the following functions tell us?

   a. =**NORM.DIST**(1080, 1059.1, $\sigma_{\bar{x}}$, 1) = 0.8656

   b. =**NORM.INV**(35%, 1059.1, $\sigma_{\bar{x}}$,) = 1051.8

Key Terms

Confidence interval for the mean	
Margin of error	
Critical value	
t-Distribution	
Degrees of freedom	

Distribution of sample proportions	

Summary

In the last objective, we started with **population** data (average SAT scores of all institutions) and looked at the new **sampling distribution** we can create by taking random **samples**. Now we want to go backwards and start with a random sample and try to make an inference about the original population. You randomly sample 15 students and ask them their age getting a sample **mean** of 20.8 years of age. From previous studies you know the population **standard deviation** is 1.2 years of age. We can estimate the true population mean age with 95% confidence assuming the population is **normally distributed**.

The **Central Limit Theorem** tells us that the **distribution of sample means** will be normally distributed and centered at the true population mean but will be less spread out:

$$\mu_{\bar{X}} = \mu \qquad \sigma_{\bar{X}} = \frac{\sigma}{\sqrt{n}} = \frac{1.2}{\sqrt{15}} = 0.31$$

So we know that there is a 95% chance our sample mean of 20.8 is within two standard errors of the mean:

$$\mu - 2 \cdot (0.31) \leq 20.8 \leq \mu + 2 \cdot (0.31)$$

We can rearrange these inequalities as shown in the video to get a 95% chance that:

$$20.8 - 2 \cdot (0.31) \leq \mu \leq 20.8 + 2 \cdot (0.31)$$
$$20.8 \leq \mu \leq 21.42$$

So we can be 95% certain our true population mean is between 20.18 and 21.42 years of age. We used our knowledge of the **95% Rule** to make an interval about the **mean** using two **standard deviations**. We can use any **percentage** we want using the **NORM.S.INV** function to give us the number of standard deviations.

Confidence Level	Alpha	Tail %	Critical Value (z_c)	
90%	10%	5%	**1.645**	=NORM.S.INV(95%)
95%	5%	2.5%	**1.960**	=NORM.S.INV(97.5%)
99%	1%	0.5%	**2.576**	=NORM.S.INV(99.5%)

For a 99% confidence interval we will have 0.5% area in each tail ($\alpha = 0.01$). We use **NORM.S.INV**(99.5%) to find the associated z-**score** = 2.576.

So we know that there is a 99% chance our sample mean of 20.8 is within 2.576 standard errors of the mean:

$$\mu - 2.576 \cdot (0.31) \leq 20.8 \leq \mu + 2.576 \cdot (0.31)$$

We can rearrange these inequalities as shown in the video to get a 99% chance that:

$$20.8 - 2.576 \cdot (0.31) \leq \mu \leq 20.8 + 2.576 \cdot (0.31)$$
$$20.00 \leq \mu \leq 21.60$$

So we can be 99% certain our true population mean is between 20.00 and 21.60 years of age. In the last example we were assuming σ was known, which along with the **Central Limit Theorem** allowed us to use z-**scores** from the **standard normal distribution** for our **confidence intervals**. If σ is unknown and assuming that our **distribution of sample means** is approximately **normal** (either $n \geq 30$ or original population is normal), then $t = \dfrac{\overline{X} - \mu}{\dfrac{s}{\sqrt{n}}}$ will

follow a t-**distribution** with $n-1$ **degrees of freedom**. The procedure is identical except we use **T.INV** instead of **NORM.S.INV** to find the **critical values**.

Confidence Level	Alpha	Tail %	Critical Value (t)	
90%	10%	5%	**1.699**	=T.INV(95%, 29)
95%	5%	2.5%	**2.045**	=T.INV(97.5%, 29)
99%	1%	0.5%	**2.756**	=T.INV(99.5%, 29)

We can similarly make estimates about polling averages because we know the **binomial distribution** is approximately normal for $n \cdot p \geq 5$ and $n \cdot q \geq 5$. Polls typically report the **percentage** or proportion of respondents who favor something rather than the number of people. So instead of a traditional **binomial experiment** where x = the number of successes, we want to look at $\dfrac{x}{n}$ = the proportion of successes.

A PEW Research poll of 1,534 adults conducted in 2016 found 48% said climate change is mostly due to human activity. The **sample** size $n = 1{,}534$ and $\hat{p} = 48\%$ so clearly $n \cdot \hat{p} \geq 5$ and $n \cdot \hat{q} \geq 5$ and our **distribution of sample proportions** will be approximately **normally distributed**, so we can use critical z-**scores**. We will use 1.96 instead of 2 (from the **95% Rule**) so you are familiar with both approaches.

95% Confidence Interval
$$48\% - 1.96 \cdot \sqrt{\frac{\hat{p} \cdot \hat{q}}{n}} \leq p \leq 48\% + 1.96 \cdot \sqrt{\frac{\hat{p} \cdot \hat{q}}{n}}$$
$$48\% - 1.96 \cdot \sqrt{\frac{0.48 \cdot 0.52}{1{,}534}} \leq p \leq 48\% + 1.96 \cdot \sqrt{\frac{0.48 \cdot 0.52}{1{,}534}}$$
$$45.5\% \leq p \leq 50.5\%$$

So we can say with 95% confidence that the true proportion of adults who think climate change is mostly due to human activity is between 45.5% or 50.5%. Alternatively the proportion is 48% plus or minus 2.5 percentage points. The margin of error is 2.5 percentage points.

*As you read, jot down notes and questions. At the end of this section's Guided Worksheets there is a space for **Reflection** and **Monitoring Your Understanding**, where you can try and answer your questions after completing the guided activities and homework.*

Notes & Questions

Guided Practice Activity #1 – Weighty Matters

The CDC publication[1], *Anthropometric Reference Data for Children and Adults, 2011-14*, contains the following table

Table 6. Weight in pounds for males aged 20 and over and number of examined persons, mean, standard error of the mean, and selected percentiles, by race and Hispanic origin and age: United States, 2011–2014

Race and Hispanic origin and age	Number of examined persons	Mean	Standard error of the mean	5th	10th	15th	25th	50th	75th	85th	90th	95th
All racial and Hispanic-origin groups[1]							Pounds					
20 years and over	5,236	195.7	0.94	136.7	146.2	154.2	165.1	189.3	218.8	236.7	249.9	275.4
20–29 years	936	186.8	2.60	126.2	137.6	143.9	152.9	177.8	208.5	232.1	247.2	280.8
30–39 years	914	198.8	1.73	140.2	150.2	158.2	168.1	190.8	221.2	242.8	259.6	281.9
40–49 years	872	201.7	1.60	146.2	156.3	162.9	171.7	196.4	222.5	237.2	249.0	279.2
50–59 years	854	199.5	2.03	140.0	152.0	160.0	170.2	195.9	222.3	236.9	250.4	279.4
60–69 years	874	199.7	3.02	137.9	147.0	154.3	168.0	195.3	223.1	244.2	255.3	279.2
70–79 years	486	189.3	2.03	136.5	146.2	152.6	166.6	183.6	212.0	227.0	236.3	251.5
80 years and over	300	174.6	1.90	125.2	132.6	141.4	154.2	171.1	194.5	207.1	216.1	233.5

1. For males 20 years and over what is the mean and median weight?

2. Why are these different?

3. Use the standard error to compute the standard deviation for males 20 years and over.

4. Construct a 95% confidence interval about the true population mean.

5. Construct a 90% confidence interval about the true population mean.

6. Which is interval is larger? Why?

[1] Fryar CD, Gu Q, Ogden CL, Flegal KM. Anthropometric reference data for children and adults: United States, 2011–2014. National Center for Health Statistics. Vital Health Stat 3(39). 2016.

<u>**Guided Practice Activity #2**</u> – Conservative Error

A June 2018 Pew Research Poll[2] of 4,587 adult Americans found that 59% feel there are too few women in high political office in the country. Interestingly, men (48%) and women (69%), as well as Republicans (33%) and Democrats (79%), show large gaps in their responses.

1. Compute the 95% margin of error for all adults using 1.96.

2. The true proportion of adult Americans thinking there are too few women in political office is between what two values with 95% confidence?

3. Compute the 95% margin of error for Republicans (assume $n = 2294$) and Democrats (assume $n = 2293$) using 1.96.

4. Note that the margin of error changes with the percentages p and q. It turns out that the margin of error is largest when both percentages are 50%. Plugging $\dfrac{1}{2}$ into the formula we get:

$$\sqrt{\frac{\hat{p} \cdot \hat{q}}{n}} = \frac{1}{2 \cdot \sqrt{n}}$$

Then using 2 for 95% confidence we get the **conservative margin of error**:

$$2 \cdot \sqrt{\frac{\hat{p} \cdot \hat{q}}{n}} = \frac{1}{\sqrt{n}}$$

Compute the conservative margin of error for this poll.

---

http://www.pewsocialtrends.org/2018/09/20/women-and-leadership-2018/

5. Women (59%, $n = 2303$) are also more likely than men (36%, $n = 2284$) to say gender discrimination is a major reason why there are not more women in high political offices. Compute the 95% confidence intervals (use 2) for these proportions using the conservative margin of error.

Excel Preview – 12.4 Binom_Test

A blood test for TB is 90% accurate.

	A	B
1		a.)
2	Number of false results	Probability
3	0	0.0%
4	1	0.0%
5	2	0.0%
6	3	0.1%
7	4	0.4%

1. If 120 college students are tested, what formula is entered in cell B3 using the COMBIN function to determine the binomial probabilities for the number of false results?

2. Compute the mean and standard deviation of this binomial distribution.

3. 95% of the data values are between what two values centered about the mean (use 2)?

4. Would it be considered unusual for 20 students to get false results?

5. What formula could be entered to compute the probability that exactly 20 students get false results using BINOM.DIST, with 0 for the last argument to get just one probability?

6. What does the following formula's output tell us?

=1 - BINOM.DIST(19,120,0.1,1) = 1.577%

Section 12.3 Reflections
Monitor Your Understanding

What surprised you in this section regarding polling?

Section 12.4 Hypothesis Testing

Objective 1 – Test Hypotheses One Sample
Objective 2 – Test Hypotheses Two Samples
Objective 3 – Test for Independence

Objective 1 – Test Hypotheses One Sample

Key Terms

Null hypothesis	
Alternative hypothesis	
t-score	
p-value	

Summary

Recall the example where we drove 30 Tesla Model S sedans and get a **sample** mean of 92 MPGe with a sample **standard deviation** of 2.7 MPGe. We computed the 90% and 99% confidence intervals shown below. In this section we want to test claims, such as: the Tesla Model S sedan gets an average of 93 MPGe. Does our sample support this claim?

90% Confidence Interval	99% Confidence Interval
$92 - 1.699 \cdot \left(\dfrac{s}{\sqrt{n}} \right) \leq \mu \leq 92 + 1.699 \cdot \left(\dfrac{s}{\sqrt{n}} \right)$	$92 - 2.756 \cdot \left(\dfrac{s}{\sqrt{n}} \right) \leq \mu \leq 92 + 2.756 \cdot \left(\dfrac{s}{\sqrt{n}} \right)$
$92 - 1.699 \cdot (0.493) \leq \mu \leq 92 + 1.699 \cdot (0.493)$	$92 - 2.756 \cdot (0.493) \leq \mu \leq 92 + 2.756 \cdot (0.493)$
$91.16 \leq \mu \leq 92.84$	$90.64 \leq \mu \leq 93.36$

Well we are 99% confident the true **mean** is between 90.64 and 93.36 so at this level of confidence we can say our result supports the claim that the true population average is 93 MPGe. But we sacrificed precision for our high level of confidence, the interval is kind of large. To be more precise we could use the 90% confidence interval from 91.16 to 92.84, but now 93 mpg is NOT in this interval.

The question we want to ask is whether or not the sample **mean** of 92 MPGe is far enough away from the claimed 93 MPGe that we can reject this claim (the null hypothesis).

Null Hypothesis (H_0): $\mu = 93$

Alternative Hypothesis (H_a): $\mu \neq 93$

As usual we measure "far away" by standardized scores, in this case **t-scores**.

$\sigma_{\overline{X}} = \dfrac{\sigma}{\sqrt{n}}$ is called the **standard error** but σ is unknown so we use $\dfrac{s}{\sqrt{n}}$ and a t-distribution.

$$ t = \frac{\overline{X} - \mu}{\dfrac{s}{\sqrt{n}}} = \frac{92 - 93}{\dfrac{2.7}{\sqrt{30}}} = -2.029 $$

Our table tells us that $t = \pm 1.699$ are the **critical values** for tails of 10% (5% in each tail). Any t-score beyond these will be in the tail and thus far away or **unusual**.

Confidence Level	Alpha	Tail %	Critical Value (t)	
90%	10%	5%	**1.699**	=T.INV(95%, 29)
95%	5%	2.5%	**2.045**	=T.INV(97.5%, 29)
99%	1%	0.5%	**2.756**	=T.INV(99.5%, 29)

Our $t = -2.029 < -1.699$ and is thus in the 10% tail, so we have enough evidence to reject the null hypothesis at $\alpha = 0.10$. We are 90% certain the claim, $\mu = 93$, is not true.

Our $t = -2.029 > -2.756$ and is thus NOT in the 1% tail, so we are 99% certain we do not have enough evidence to reject the null hypothesis at $\alpha = 0.01$.

Is $t = -2.029$ far enough away from the mean for us to reject Elon Musk's claim? It can be confusing that we answer: **"Yes, reject at $\alpha = 10\%$.",** but **"Not enough evidence to reject at $\alpha = 1\%$."** The idea is that we assume Elon is innocent until proven guilty. We need the evidence to make us sure beyond "reasonable doubt," but this is of course subjective! So we can be 90% certain the true mean is not 93 MPGe, but not 99% certain. The default for "reasonable doubt" is 95% certainty.

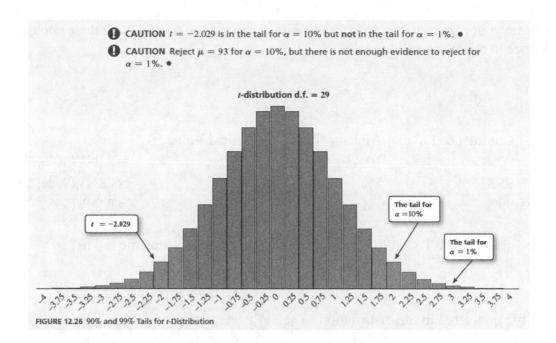

CAUTION $t = -2.029$ is in the tail for $\alpha = 10\%$ but **not** in the tail for $\alpha = 1\%$. ●

CAUTION Reject $\mu = 93$ for $\alpha = 10\%$, but there is not enough evidence to reject for $\alpha = 1\%$. ●

t-distribution d.f. = 29

$t = -2.029$

The tail for $\alpha = 10\%$

The tail for $\alpha = 1\%$

FIGURE 12.26 90% and 99% Tails for t-Distribution

As you read, jot down notes and questions. At the end of this section's Guided Worksheets there is a space for **Reflection** *and* **Monitoring Your Understanding**, *where you can try and answer your questions after completing the guided activities and homework.*

Notes & Questions

Guided Practice Activity #1 – Buying Houses

A realtor claims the mean price of a home is at least $185,000. You sample 8 such homes and compute a sample mean of $178,425 with a standard deviation of $7,450. Test the claim at $\alpha = 0.05$.

1. The claim is that the average price of these homes is greater than or equal to $185,000. We want to know if our sample mean of $178,425 is far enough **below** the mean to allow us to reject the claim, so we put all 5% of the area in the left tail. What are the null and alternative hypotheses?

2. Compute the *t*-score for your sample.

3. Compare your *t*-score to the appropriate critical value from the table. Is there enough evidence to reject the null hypothesis?

Confidence Level	Alpha	Tail %	Critical Value (*t*)	
90%	10%	10%	-1.415	=T.INV(90%, 7)*-1
95%	5%	5.0%	-1.895	=T.INV(95%, 7)*-1
99%	1%	1.0%	-2.998	=T.INV(99%, 7)*-1

4. The *p*-value =**T.DIST**(-2.496, 7, 1) = 2.06%. What does this statistic tell us?

5. Identify you t-score in the distribution. The 5% tail is patterned.

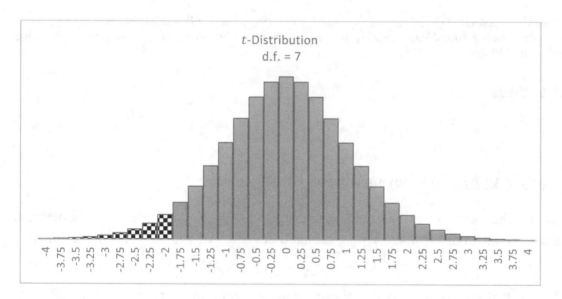

6. How do we use the *p*-value to test the hypothesis?

Key Terms

Variance	
Statistically significant	

Summary

In the last section we worked with a single **sample** of data and tested a claim about the **population** mean. In this section we want to compare the means between two samples. Comparing groups is a fundamental part of experimental design. Typically one group receives an intervention (study guide, new medicine, etc.) and we measure results for this group as well as a control group that did not receive the intervention.

There are two biology classes being taught, Class A uses active learning strategies and Class B does not. You want to test if the scores on the shared final exam are different.

	$\overline{X}$	s	n
Class A	84.0	9.99	35
Class B	78.3	13.88	42

Use the data given in the *Class_t-test* sheet in **Data Sets**, assuming the **variances** are not equal to test the claim that the scores are different at $\alpha = 0.05$.

Null (H_0): $\mu_1 = \mu_2$

Alternative (H_a): $\mu_1 \neq \mu_2$

We are going to run the two sample *t*-test assuming unequal variances.

t-Test: Two-Sample Assuming Unequal Variances		
	Variable 1	Variable 2
Mean	84.01174	78.26093
Variance	99.85611	192.5709
Observations	35	42
Hypothesized Mean Difference	0	
df	74	
t Stat	**2.108627**	
P(T<=t) one-tail	0.019181	

t-Test: Two-Sample Assuming Unequal Variances	
t Critical one-tail	1.665707
P(T<=t) two-tail	**0.038362**
t Critical two-tail	**1.992543**

It's that easy! Now all of our hard work in the previous sections pays off. We identify our **t-score**: $t = 2.109$ and the critical cutoff value for a two tailed test: $t = 1.993$. Our *t*-score is greater than the **critical value**, thus we are in the 5% tail and we reject the null hypothesis $\mu_1 = \mu_2$. So we can say that there is a **statistically significant** difference between the scores at alpha $= 0.05$.

Another way for us to proceed is to use the **p-value**: $p = 0.038$, which represents the **probability** that a **sample statistic** (*t*-score) will be 2.109 or greater in either tail:

$P(|t| \geq 2.109) = 0.038 = 3.8\%$. Since this probability is less than 5% it means our *t*-score is in the tail, and we can reject the **null hypothesis**.

CAUTION! If $p \leq \alpha$ then reject H_0. This can be a quick way to test your hypothesis and many people report *p*-values when running statistical tests, not critical values.

*As you read, jot down notes and questions. At the end of this section's Guided Worksheets there is a space for **Reflection** and **Monitoring Your Understanding**, where you can try and answer your questions after completing the guided activities and homework.*

Notes & Questions

Guided Practice Activity #1 – Braking (excel Preview 12.7)

We are given the braking distances of two types of tires in feet; and want to test the hypothesis that the mean distance is zero.

	X_bar	s	n
Tire A	38.9	4.77	35
Tire B	41.7	5.22	28

1. We run an F-test (as shown in e-text) for equal variances. Which tire is variable 1?

F-Test Two-Sample for Variances		
	Variable 1	Variable 2
Mean	41.70757011	38.87935
Variance	27.20295526	22.79333
Observations	28	35
df	27	34
F	1.193461065	
P(F<=f) one-tail	0.309776775	
F Critical one-tail	1.816691026	

2. What is the p-value for the F-test? Is it less than 5%? Are the variances equal or unequal?

3. What is the t-score for the t-test? What is the critical value? Is our t-score in the tail?

t-Test: Two-Sample Assuming _____ Variances		
	Variable 1	Variable 2
Mean	41.70757011	38.87935
Variance	27.20295526	22.79333
Observations	28	35
Pooled Variance	24.74513291	
Hypothesized Mean Difference	0	
df	61	
t Stat	2.242388648	
P(T<=t) one-tail	0.014291038	
t Critical one-tail	1.670219484	
P(T<=t) two-tail	0.028582077	
t Critical two-tail	1.999623585	

4. Should we reject that the tires have the same mean braking distance?

5. What is the p-value for the t-test? Is it less than 5%? Should we reject that the tires have the same mean braking distance?

Objective 3 – Test for Independence

Summary

In this last section we want to show you one more type of test involving two-way or contingency tables.

Given a **sample** of 46 people, we know whether or not they attended a webinar on solar panels and whether or not they actually installed solar panels (as shown in the observed values table). We want to test whether or not there is a relationship between attending our webinar and installing the solar panels.

Null (H_0): Installing solar panels is independent of attending the webinar.

Alternative (H_a): Installing solar panels depends on attending the webinar.

▲	A	B	C	D	E	F	G	H	I	J	K	L
1		Observed Values						Expected Values				
2			Attended Webinar						Attended Webinar			
3				No	Yes	Total				No	Yes	Total
4		Installed Solar	No	22	11	33		Installed Solar	No	20.09	12.91	33
5		panels	Yes	6	7	13		panels	Yes	7.91	5.09	13
6			Total	28	18	46			Total	28	18	46
7												
8												
9		p-value:	0.199294									

The only tricky part of running a chi-square test for independence is determining the expected values. We know that 33 out of 46 people did not install panels (71.7%) and 28 out of 46 people did not attend the webinar (60.9%). Assuming these variables are **independent** the **probability** of someone not attending and not installing is 71.7% * 60.9% = 43.7%. We therefore EXPECT **43.7% * 46 = 20.09** people in the No/No box.

CAUTION! We need all the expected values to be greater than 5 to use a chi-square test. Once we compute all the expected values, we simply use:

$$=\textbf{CHISQ.TEST}(D4:E5,J4:K5) = 0.1992$$

The question at hand is whether or not the observed values are far enough away from the expected values. Yes it is the same question over and over! The output of the function is our friend the *p*-value. In this case:

$$p = 0.1992 > \alpha = 0.05$$

Thus the observed values are NOT that far away, they are not in the tail; so we do NOT have enough evidence to reject the **null hypothesis** that installing solar panels is **independent** of attending the webinar. Tough luck for us! We were hoping these variables were related, meaning there was some effect our webinar had on actually installing panels.

*As you read, jot down notes and questions. At the end of this section's Guided Worksheets there is a space for **Reflection** and **Monitoring Your Understanding**, where you can try and answer your questions after completing the guided activities and homework.*

Notes & Questions

Guided Practice Activity #1 – Smoking Sex

We are given the number of times a sample of people have tried to quit smoking and their gender. At alpha = 0.05 can you infer that the number of times they have tried to quit is related to gender?

1. Compute the row and column sums in the contingency table.

Observed Values					
	Number of times tried to quit smoking				
		1	**2..3**	**4+**	**Total**
Gender	**Male**	265	248	120	
	Female	163	112	85	
	Total				

2. What percentage of the total are men?

3. What percentage of the total have tried to quit once?

4. What percentage and number of the total are men and have tried to quit once?

5. Fill in the expected values:

Expected Values					
	Number of times tried to quit smoking				
		1	**2..3**	**4+**	**Total**
Gender	**Male**	☐	☐	☐	☐
	Female	☐	☐	☐	☐
	Total	☐	☐	☐	☐

6. We compute the p-value using CHISQ.TEST and get 0.028. Is your *p*-value less than 5%? Is the number of times former smokers tried to quit independent of gender?

Excel Preview – 12.6 Burrito

In this problem we explore the concepts of hypothesis testing and confidence intervals. Your local vegan bistro advertises a burrito that has 750 calories. You want to test this so you have people randomly arrive throughout the week and buy 12 burritos. You bring them to the lab and measure the caloric content by burning each one (not eating them!) in a calorimeter and measuring the temperature rise in water. A Calorie raises 1 kg of water by 1 degree Celsius. You get an average of 753.7 Calories with a standard deviation of 6.11 Calories.

	I	J	K	L	M
		Mean:	753.7		
		Std dev.:	6.11		
a.)		**Two Sided Test Critical Values**			
		Confidence Level	Alpha	Tail %	Critical Value (t)
		90%	10%	5%	1.796
		95%	5%	2.5%	2.201
		99%	1%	0.5%	3.106

1. To compute the critical t-values for the three confidence intervals using **T.INV(probability, deg_freedom)**, what function is entered in cell M7?

2. Compute the standard error and the t-value for our sample mean.

3. Determine if you can reject at each confidence level by comparing your t-value to the critical values. If you are in the tail then reject! Answer yes or no.

4. To compute the p-value, we can use **T.DIST(x, deg_freedom, cumulative)** which will give you the percentage of t-scores less than your t-value, so subtract this from 1 to get the percentage in the right tail, and multiply by 2 to get area in both tails. What formula can we type into Excel?

5. If $p = 5.98\%$ (from Excel no rounding) how do we use it to test the hypothesis at each alpha?

Section 12.4 Reflections
Monitor Your Understanding

This brings us to the end of our journey into **statistics**. Let's end with a favorite definition. Rejecting the null hypothesis is said to be significant because we are 5% sure it could not have occurred by chance.
Definition: A result is **statistically significant** if it is unlikely to have occurred by chance. You reading this book is unlikely to have occurred by chance and thus is significant!

What have you found most significant in this chapter?

Chapter 13 – Modeling the Real World

Section 13.1 Modeling Basics

Objective 1 – Understand Parameters and Constraints
Objective 2 – Explore a Conceptual Model

Objective 1 – Understand Parameters and Constraints

Key Terms

Model	
Parameter	
Constraint	
Occam's razor	

Summary

All of us use **models** every day to help make decisions from where to eat, to what car to buy, to who we spend time with, and how much we exercise. Cathy O'Neill, in her fantastic book, *Weapons of Math Destruction*, defines a model as "nothing more than an abstract representation of some process" that takes what we know and uses this information to predict responses in various situations and help us make informed decisions. Anthony Starfield in his book, *How to Model It: Problem Solving for the Computer Age*, says that a model is "a caricature of the real world" and reminds us of two important guidelines to consider when creating any model:
1. What is the purpose of the model?
2. What are the **constraints** of the model?

The map model in Example 1 in the text illustrates how a model needs to simplify the complex real world and strip away inessential information. The principle of making a minimum number of assumptions in creating a model is referred to as "**Occam's razor**". Our original map model does not account for wind speed, potholes, restaurant options, or the chance of hitting a moose. Also note that our model did not have any formulas or equations, since the purpose was choosing a route. If you were modeling the cost of an entire weeklong trip to Maine from your home, then formulas and use of a spreadsheet would be crucial. Assumptions are clearly built into this second model, the $0.55 per mile rate and rental car mpg are treated as constants but we could vary them.

The power of a **model** is the ability to ask "What if?" questions. What if we can get plane tickets for $125 or the rental car is going to be $400 for the week? Should you fly or drive? Our model doesn't answer that question! It allows you to tweak the assumptions and inputs and **iterate** over different scenarios, generating insights that help you make an informed decision.

As you read, jot down notes and questions. At the end of this section's Guided Worksheets there is a space for **Reflection** *and* **Monitoring Your Understanding***, where you can try and answer your questions after completing the guided activities and homework.*

Notes & Questions

Guided Practice Activity #1 – Maine Trip

Use the spreadsheet to answer the following:

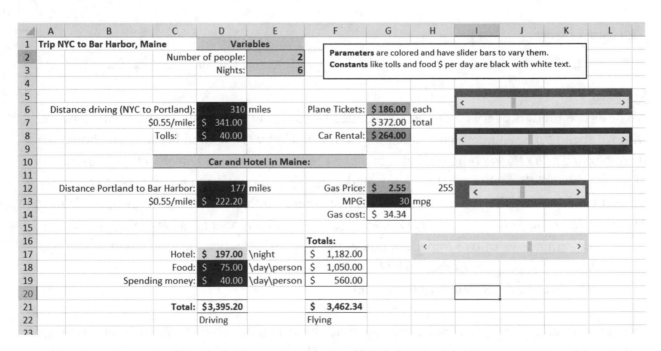

1. What is the purpose of this model?

2. What are the constraints?

3. This worksheet is printed black and white, making the parameters harder to identify. List all the parameters. How do you vary them?

4. Why are these parameters and not variables since they vary?

5. Which cells have formulas in them?

6. What formulas are entered in each of them? Note: E2 has been named "peeps" and E3 "nights", costs are roundtrip, 50 miles extra driving in Bar Harbor is total not round-trip, and think about relationship between nights and days.

Objective 2 – Explore a Conceptual Model

Key Terms

Iteration	
Conceptual Model	

Summary

Let's look at another type of **model** that involves just an equation. We can create a model that will estimate the time required to do your homework in this course each week. We assume that you will be asked to read so many pages (p) for a certain number of sections (s), each of which has a Mastery Check quiz. There will be a certain number (n) of MyMathLab problems, and a certain number (e) of Excel problems. There are four different times in this model:

1. Time to read one page (t_p).
2. Time to complete one Mastery Check quiz (t_q).
3. Time to do one MML problem (t_M).
4. Time to do one Excel problem (t_e).

Thus a simple model for total time is:

$$\text{Time}_1 = p \cdot t_p + s \cdot t_q + n \cdot t_M + e \cdot t_e$$

Modeling the world around us is one of the fundamental aims of science, which is the systematic study of the structure and behavior of the physical and natural world through observation and experiment. Climate scientists in particular spend enormous effort creating and refining models of the incredibly complex interactions driving (forcing) climate change, both natural and human caused (anthropogenic). The examples so far have introduced basic concepts of what a model is and how it can help us make decisions.

As you read, jot down notes and questions. At the end of this section's Guided Worksheets there is a space for **Reflection** *and* **Monitoring Your Understanding**, *where you can try and answer your questions after completing the guided activities and homework.*

Notes & Questions

Guided Practice Activity #1 – Doing Your Homework

We can make a more sophisticated model by assuming that more time spent on the reading will lead to less time spent on everything else. We can introduce a reading **variable** (r) that represents the multiple of our usual reading time (t_p).

1. Introduce the reading variable into the simple equation in the summary above so that our time reading is a factor of it. Call this equation *Time_1.2*.

2. Now we want to make the assumption that more reading reduces the time spent on everything else. So if we spend twice the time reading the other times are cut in half, three times are times are a third etc. Write down the new equation and call it *Time_2*.

3. What if instead of taking fractions of the times we take the rth root of the times? What is the equation for this situation? Call it *Time_2.5*.

4. Would the rth roots equation give lower or higher times than Time_2?

5. In the text we make the assumption that more reading would affect the quiz time the most and we divide that by r^2. You decide that the Excel problems' time would not be reduced so much (half, third etc.). How can you reduce the Excel problem time but not as much as in *Time_2*? Write down a new equation that incorporates the new quiz and Excel times, call it *Time_3.5*.

Excel Preview – 13.1 Car

You are buying a car and wish to do a cost/benefit analysis on buying a more expensive Prius, which gets better mileage, $\dfrac{50 \text{ city}}{54 \text{ highway mpg (miles per gallon)}}$, than a less expensive Subaru

Impreza, which gets $\dfrac{28 \text{ city}}{38 \text{ highway mpg}}$. The spreadsheet is showing outputs using full decimal place values of inputs. The website, https://newsroom.aaa.com/auto/your-driving-costs/ , also tells us the average annual maintenance cost and annual depreciation (amount the car loses in value each year).

	A	B	C	D	E
1		Impreza	Prius		
2	Cost	$ 18,995.00	$ 28,476.00		
3	Maintenance:	$ 1,250.00	$ 900.00		
4	Depreciation:	$ 2,114.00	$ 5,704.00		
5					
6					
7				Impreza	Prius
8				a.)	
9				Gallons per 100 miles	
10	MPG Hway	38	50	2.63	2.00
11	MPG City	28	54	3.57	1.85
12				b.)	
13				Gallons Used	
14	Miles Hway	15,000		394.7	300.0
15	Miles City	7,500		267.9	138.9
16				c.)	
17			Total Gallons:	662.6	438.9
18	Price per Gallon	$ 2.80	Cost:	$ 1,855.26	$ 1,228.89

1. What parameters are in this screen shot?

2. To compute the number of gallons each car uses to travel 100 miles on the highway and 100 miles in the city, what formula would we enter in cell D10 that could be filled down and across?

3. To compute the number of gallons each car uses to travel the miles given for highway and city, what formula would we enter in cell D14 that could be filled down and across?

4. In this situation how long will it take to recoup the extra money you spent on the Prius given the annual gas and maintenance savings?

5. How does the depreciation impact your decision on which car to buy?

<u>Section 13.1 Reflections</u>
Monitor Your Understanding

What surprised you in this section on modeling?

| **Objective 1** – Understand Rate Models |
| **Objective 2** – Explore Logistic Models |

| **Objective 1** – Understand Rate Models |

Key Terms

Heuristic	

Summary

The **models** we study in this section will look at processes that change over time, like populations, the spread of rumors, or the temperature of a cooling cup of coffee. One of the hallmarks of a good modeler is being able to recognize when a given situation falls into a known modeling category, which has known "rules of thumb" or **heuristics**. Some popular heuristics are:

- Rephrase or explain the problem to someone else.
- Draw a diagram.
- Look for simplifying assumptions and key **variables**.
- Choose effective notation and build a formula.
- Try to imagine the solution as a graph or table or equation.

Some basic notation will help in our analysis of population models:

P_n = population n years since initial value

P_{n+1} = population $n + 1$ years since initial value

$P_{n+1} - P_n$ = annual change in population in year $n + 1$

The change in population is key to these rate models. Simplifying assumptions such as fixed population growth a year gives linear growth, while assuming the annual change is proportional to the population at the start of the year gives exponential growth. Introducing a harvesting rate for wildlife populations can keep the population in check.

As you read, jot down notes and questions. At the end of this section's Guided Worksheets there is a space for **Reflection** *and* **Monitoring Your Understanding**, *where you can try and answer your questions after completing the guided activities and homework.*

Notes & Questions

You are working for the Fisheries and Wildlife Department and are restocking a stream that has been restored to health after decades of industrial pollution. Assume you release 1,000 brook trout. Model the population growth with sustainable harvesting (fishing).

1. Assuming the fish population grows by 75 fish per year, write down an equation modeling the population.

2. Assuming the fish population grows by 7.5% per year write down the equation modeling the population.

3. How can a harvesting term of 100 fish per year be added to these models?

4. Which of these models with harvesting 100 fish per year would yield the following scatterplot?

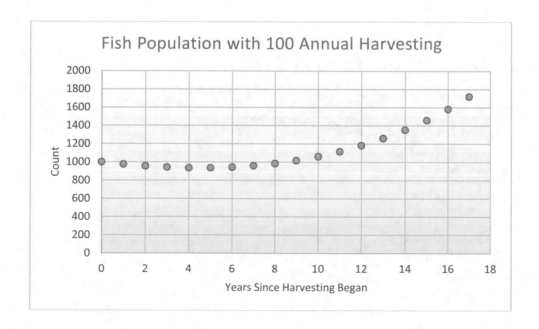

5. How can a harvesting term that starts at 50 and grows by 3% per year be added into these models?

6. What are the formulas that are entered in cells **B3** and **C3** for the models with a harvesting term that starts at 50 and grows by 3% per year?

	A	B	C
1	Years Since	Linear	Exponential
2	0	1000	1000
3	1	1023.5	1023.5
4	2	1045.5	1047.2

7. Which of these models with a harvesting term that starts at 50 and grows by 3% per year would yield the following scatterplot?

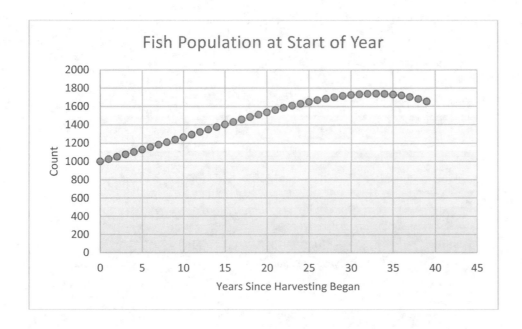

Objective 2 – Explore Logistic Models

Key Terms

Carrying capacity	
Logistic model	

Summary

The fish population **model** assumed the growth per unit time was **proportional** to the existing population. This led to a constant annual **growth rate** and an **exponential** model, which is a standard way to think about population growth. An obvious limitation to using exponential growth is that eventually the population will become HUGE (this is a principle hallmark of exponential growth), and the environment will not be able to support the population. There will become so many fish they will have to be stacked on top of each other! We introduced harvesting to control the growth, another option is to introduce the **carrying capacity**, or maximum population the environment can support.

A version of this type of problem is the spread of a disease or rumor. We cannot infect more than 100% of the population so this effectively becomes a **carrying capacity**. To create a model that will estimate the time required for a rumor to spread through an entire population, we need to make some basic assumptions to get started:
1. Assume initially a small percentage of the population has heard the rumor.
2. Assume an initial **growth rate** that is high for the spread of the rumor.
3. Make this growth rate decline to zero as the number infected goes to 100%.

It is easy enough to assume we start with 2% infected and a **growth rate** of 30%. The tricky part is making the growth rate go to zero as the number infected goes to 100%. Can you think of how to make the new growth rate decline to 0%?

▲	A	B	C	D	E	F
1	Growth Rate:		30%			
2	Carrying Capacity:		100%			
3		Hour	Current Number of Infected		New Growth Rate	New Number of Infected
4		0	2%	98.00%	29.4%	2.59%
5		1	2.59%	97.41%	29.2%	3.34%
6		2	3.34%	96.66%	29.0%	4.31%
7		3	4.31%	95.69%	28.7%	5.55%

We can use the fact that the number goes to 100% to create a formula that takes a fraction of the 30%:

$$\text{rate}_{new} = \frac{100\% - \text{percentage infected}}{100\%} \cdot 30\%$$

This will give us a new growth rate smaller than 30% and approaching 0% as the number infected goes to 100%.

CAUTION! This formula is not obvious, an important part of becoming a modeler is trying to come up with solutions before reading them.

Running the model for 30 hours we can see in the scatterplot that the number infected is basically 100% (99.67%). It is not obvious that this **model** will never exceed the **carrying capacity** but mathematicians have studied this model extensively and shown that the limit of the number infected is 100%.

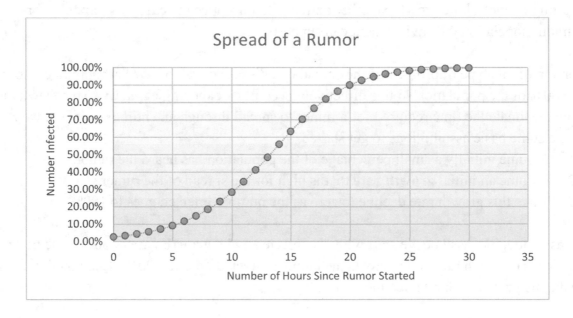

This *s*-shaped curve is called a **logistic model** and as mentioned above was first explored related to population growth. The carrying capacity will be the limiting population value no matter what population you start with or initial **growth rate** you select.

As you read, jot down notes and questions. At the end of this section's Guided Worksheets there is a space for **Reflection** *and* **Monitoring Your Understanding***, where you can try and answer your questions after completing the guided activities and homework.*

Notes & Questions

– Fish Capacity

We can use the idea of carrying capacity to create a spreadsheet model for the fish population.

	A	B	C	D	E	F
1	Growth Rate:		7.5%			
2	Carrying Capacity:		15,000			
3		Year	Current Number of Fish	(Capacity - Current)/Capacity	New Growth Rate	New Number of Fish
4		0	1000.00	0.93	7.0%	1070.00
5		1	1070.00	0.93	7.0%	1144.53
6		2	1144.53	0.92	6.9%	1223.82
7		3	1223.82	0.92	6.9%	1308.11

1. We enter the growth rate, carrying capacity and initial fish population. What formula is in cell **D4**?

2. How is this used to create the new growth rate in cell **E4**?

3. Why are we guaranteed that the new growth rate will go to 0%?

4. What formula is in cell **F4**?

5. What kind of curve is shown in the scatterplot?

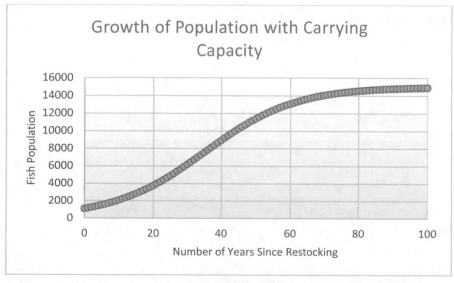

6. Why do you think this graph is sometimes called a "learning curve"?

7. What is the limiting (maximum) value of the fish population?

Excel Preview – 13.4 Pop_1

A simple population model has you split the population into two groups: adults and young. Then each year a percentage of adults survive and a percentage of young mature into adults. The new number of young will be a percentage of the adults, which can be over 100%.

◢	A	B	C	D	E	F	G
1							Rates
2		a.)				Survival:	75%
3		Generation	Adults	Young		Maturity:	35%
4		1	100.0	40.0		Birth:	60%
5		2	89.0	60.0			
6		3	87.8	53.4			
7		4	84.5	52.7			

1. Given 100 adults and 40 young; and a survival rate of 75% for the adults, a maturity rate of 35% for the young and a birth rate of 60%; what formulas for the number of adults and young in the second generation can be entered and filled down?

2. Which of the percentages can be over 100%? Explain!

3. The current situation has the population declining as shown in the scatterplot. If we keep the survival and birth rates the same, how high do you think the maturity rate has to be for the populations to increase?

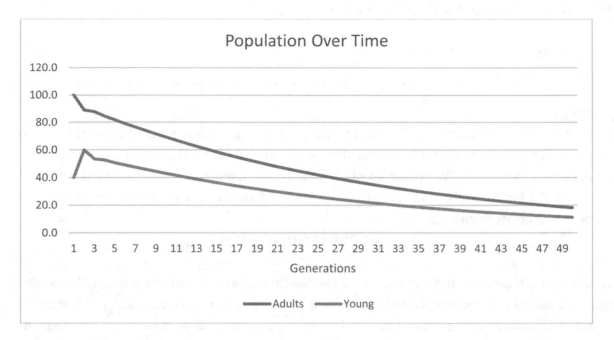

4. Finally if this is a model for fish populations, what do you think would be appropriate for the survival, maturity, and birth rates?

Section 13.2 Reflections
Monitor Your Understanding

What was most interesting about the population models?

| Objective 1 – Explore Random Simulations |
| Objective 2 – Explore Queueing Model |

| Objective 1 – Explore Random Simulations |

Key Terms

| Stochastic Model | |

Summary

In Chapter 10 we introduced the **RAND** function and used it explore random simulations. We will apply **RAND** to some of our models already introduced. RAND will generate a random number between 0 and 1. We can multiply RAND by 10 to generate a random (real with decimals not whole) number between 0 and 10, and we can subtract 4 from this to generate a random number between -4 and 6.

$$-4 \; < \; 10*\textbf{RAND}() - 4 \; < \; 6$$

Models with randomness built-in are called **stochastic models**. We create a new fish population stochastic model that incorporates randomness into the annual **growth rate** and the annual harvest. Let's consider the harvest rate of 150 fish/year first. This parameter can be made to randomly change each year from anywhere between 80% and 120% of 150. This would seem to capture some of the inherent variability in how many fish are actually caught, based on the number of licenses we sell. To accomplish we use the **RANDBETWEEN** function.

RANDBETWEEN will generate an integer (whole number plus and minus) between two given numbers. So we can use this to generate a random whole percentage between 80% and 120%:
$$80\% \; < \; \textbf{RANDBETWEEN}(80, 120)/100 \; < \; 120\%$$

Then multiply this by the catch which is currently set to 150 fish/year to generate a random catch between 120 and 180:
$$\textbf{120} \text{ fish/year} \; < 150*\textbf{RANDBETWEEN}(80, 120)/100 \; < \; \textbf{180} \text{ fish/year}$$

Next we want to randomize the annual growth rate of 7.7%. The variability of this parameter seems more volatile than the catch per year so will use slider bars to allow for changing the lower and upper bounds of the **RANDBETWEEN** function; instead of just setting them to fixed numbers like 80 and 120 above.

The lower bound will be allowed to range from 0% to 100% while the upper bound will be allowed to range from 100% to 500%.
$$0 < \text{Lo} < 100 \quad \text{and} \quad 100 < \text{Hi} < 500$$

Thus the full range of the **growth rate** (currently 7.7%) will be:

$$-7.7\% \;<\; 7.7\%*\textbf{RANDBETWEEN}(\,-Lo,\,Hi)/100 \;<\; \textbf{38.5\%}$$

CAUTION! The slider bars only accept positive range so we introduce a negative sign in **RANDBETWEEN**.

As you read, jot down notes and questions. At the end of this section's Guided Worksheets there is a space for **Reflection** *and* **Monitoring Your Understanding***, where you can try and answer your questions after completing the guided activities and homework.*

Notes & Questions

Guided Practice Activity #1 – Random Fish

The spreadsheet shows the stochastic model for the fish population created in the text.

J2			f_x	=H2+I2-catch*(RANDBETWEEN(80,120)/100)				

	A	B	C	D	E	F	G	H	I	J
1							Year	Population at Start of Year	Population Growth	New Population at End of Year
2	Growth Rate:	Bounds:	<		>		0	1250	288	1415
3	7.7%	50					1	1415	88	1326
4		300	<		>		2	1326	-48	1145
5	Harvest:						3	1145	78	1081
6	150						4	1081	191	1143
7							5	1143	185	1208
8							6	1208	255	1320

1. We can see the formula in cell J2 in the formula bar. Explain what each term in the formula is doing.

2. The population growth uses a random percentage of 7.7%. Explain each term in the formula in cell **I2: =Rate*(RANDBETWEEN(-Lo, Hi)/100)*H2**

3. With the bounds set to 50 and 300, what are the current bounds for the growth rate?

4. The following scatterplot was created with a **Lo** bound = 50 and a **Hi** bound = 238. Explain why the population of fish collapses.

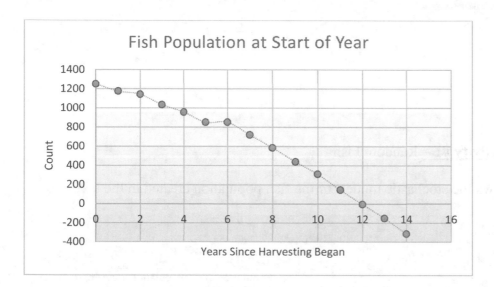

5. How could we adjust the model using a **Lo** bound = 50 and a **Hi** bound = 238 to ensure the population does not collapse (as shown below)?

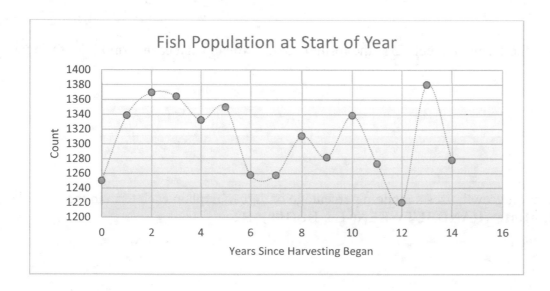

Key Terms

Queueing Model	

Summary

There are many situations where people have to wait in line (or queue) for service. An important question is how many employees need to be hired to process the customers in a timely fashion, but not so many you are wasting money paying employees to be idle.

Assume you are running a business that has a help line. You know that on average 15 customers call per minute and your employees can service calls at an average rate of 2 calls per minute. Determine how many employees you need to hire so that the wait time for customers is not annoying. The problem is incredibly simple to state yet the potential complexity is daunting. What does "on average 15 customers call per minute" mean? If we are guaranteed exactly 15 per minute and we are guaranteed to serve exactly 2 calls per minute then it seems clear we should hire 8 people and have a wait time of less than 30 seconds. But what if 20 people call in and we are only serving 1 call per minute? It's possible this can happen…

At this point you should stop and think about how you would model this situation.

The key to modeling this problem is going to be capturing the variability implicit in the "on average" statements. We will need to quantify how likely it is that 20 people call in a given minute, or that we are only serving 1 call per minute in a given minute. This will require a working knowledge of **binomial distributions** from Chapter 12 and the logic functions from Chapter 10. The argument will be accessible to anyone and the video will provide details.

Let's begin with the **distributions** and then work backwards to how we arrived at them. We are told that "on average" we serve 2 calls per minute, and see that the service rate of 2 calls per minute has the highest probability of 27.5%, but it is also very likely (27.1%) in any given minute we are only serving 1 call per minute. There is a 3.5% chance we are cranking through 5 calls per minute, but basically zero chance of serving 9 calls or more.

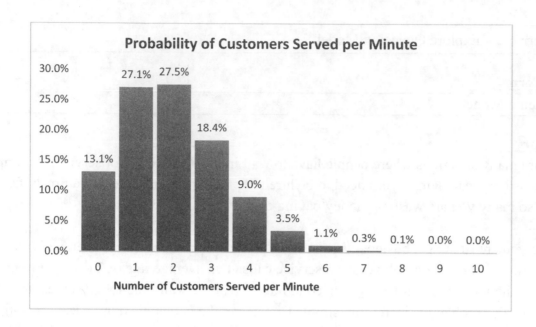

Probability of Customers Served per Minute

This is a lot of detailed information from the given "on average 2 calls per minute"! The idea is that if we expect 2 calls per minute, then in every second we would expect a $\frac{2}{60} = \frac{1}{30}$ chance of serving a call. So we can consider the 60 seconds in a minute as $n = 60$ trials, each of which has a **probability** of success: $p = \frac{1}{30}$. This is a classic **binomial experiment** with **binomial distribution** of probabilities shown in the histogram.

Similarly, if we are expecting 15 calls per minute on average, then there is a $\frac{1}{4}$ chance of getting a call in any given second, so we can consider the 60 seconds in a minute as $n = 60$ trials, each of which has a **probability** of success: $p = \frac{1}{4} = 25\%$.

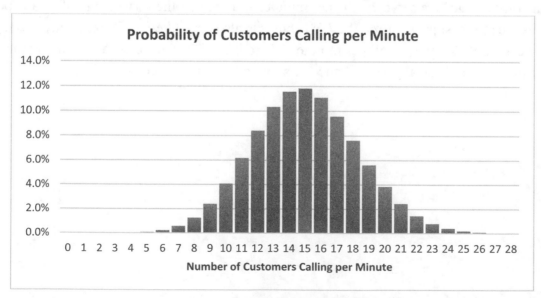

Probability of Customers Calling per Minute

Now that we have quantified how likely it will be to get 20 calls (3.8%) and serving 1 call per minute (27.1%) we want to use these distributions to generate a random number of callers and service rates every minute.

Service Time:				Employees:	
2 customers/minute				9	
3.3% one customer/sec				18	avg cust/min

Minutes	New Customers	Total Customers	Service Rate	Processed Customers	Customers on Hold
1	15	15	18	15	0
2	13	13	9	9	4
3	23	27	18	18	9
4	11	20	0	0	20

This spreadsheet model is generating a random number of callers in the **New Customers** column and a random service rate in the **Service Rate** column which is then multiplied by the number of employees to obtain the total number of customers our employees can process that minute. So if we have 9 employees we would expect to serve 18 customers per minute assuming a rate of 2 calls/minute.

As you read, jot down notes and questions. At the end of this section's Guided Worksheets there is a space for **Reflection** *and* **Monitoring Your Understanding**, *where you can try and answer your questions after completing the guided activities and homework.*

Notes & Questions

Guided Practice Activity #1 – Call Center

We will make sense of the spreadsheet for the call center model.

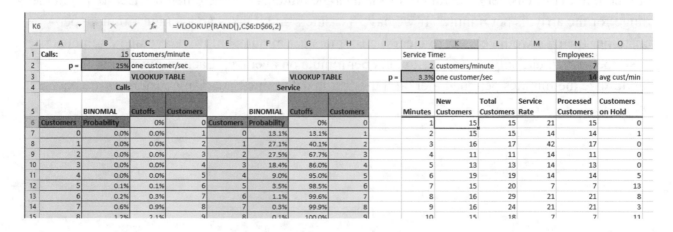

1. Note the formula bar is showing the =**VLOOKUP(RAND(), C$6:D$66, 2)** formula in cell **K6**. The binomial probabilities for the call distribution above are in column **B**, and they have been turned into the cutoffs for the **VLOOKUP** table in column **C**. Explain the 0.6% in **B14** and 0.9% in **C14**.

2. What does the **VLOOKUP** function in **K6** do?

3. The service rate in **M6** also uses a **VLOOKUP** function, but now uses the table **G6:H66**. What is the formula in cell **M6** giving the total number of customers our employees can serve in that minute?

4. What is the formula for the *Total Customers* in cell **L7**?

5. What is the formula for the *Processed Customers* in cell **N7**?

Excel Preview – 13.6 Check-out

A model of the wait at check-out can be made using just **RANDBETWEEN**. The parameters are the number of cashiers, the range for the average minutes per customer at check-out, and the range for the arrivals per minute. If we have 8 cashiers and it takes an average 1 minute per customer then we would expect to clear 8 customers per minute, but if we average 2 minutes per customer we would only clear 4 customers per minute on average. Each minute we will randomize the minutes per customer, and use this to determine how many customers are checked out.

	A	B	C	D	E	F
1						
2		Number of Cashiers:		4		
3		Minutes per Customer:		1	to	4
4			Arrivals:	0	to	5
5			a.)	b.)	c.)	d.)
6		Minutes	Arrivals	Total	Departures	Back-up
7		1	2	2	2.00	0.0
8		2	0	0.0	1.33	0.0
9		3	3	3.0	4.00	0.0
10		4	3	3.0	1.33	1.7
11		5	3	4.7	1.33	3.3
12		6	2	5.3	1.33	4.0

1. What function is in cell **C7**?

2. What function is in cell **D7**?

3. What function is in cell **E7**?

4. What function is in cell **F7**?

5. How does the total in **D8** differ from **D7**?

6. The current situation results in the following scatterplot in one iteration? Do you think 4 cashiers is viable?

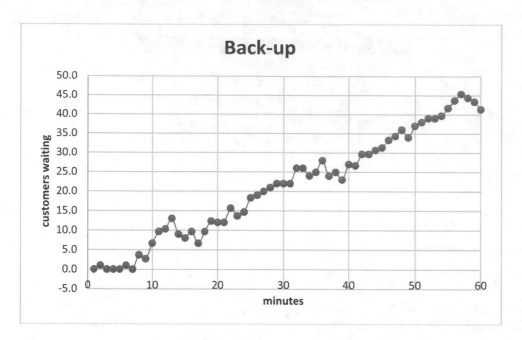

7. How many cashiers would you guess are optimal? Create the spreadsheet to check your answer!

Section 13.3 Reflections
Monitor Your Understanding

What are the constraints of the cashier model?

Section 13.4 Linear Programming (Optimization) Models

Objective 1 – Work with Constraints
Objective 2 – Use Solver

Objective 1 – Work with Constraints

Key Terms

Linear programming model	
Feasible region	

Summary

In this section we study a class of problems that are trying to optimize something, either maximizing profit, productivity, score etc. or minimizing cost, effort, space etc. There is an entire branch of mathematics, operations research, that has developed the study of such problems.

You have a small company that builds either tables or cabinets. A table costs you $300 per unit, takes 5 employee hours per unit, and generates $400 in profit per unit. Cabinets cost you $400 per unit, take 20 employee hours per unit, and generate $1,000 in profit. You are **constrained** by a weekly budget of $6,000 and 200 employee hours for your five workers. Determine the optimal combination of tables and chairs you should build weekly to maximize the profit function. Obviously building cabinets generates tremendous profit but each cabinet takes 4 times as long to build and takes up more of your weekly budget. The interplay of costs and benefits in these problems make them perfect for modeling.

Your first step in any given situation is to understand the problem, and the relationships between the variables and constraints. You might start by considering how many of just one item you can build subject to the constraints. At most we can build 20 tables at a cost of $300 each which maxes out our budget and generates $8,000 in weekly profit. Note we only use up 100 of our allowed 200 employee weekly hours.

Tables	Cabinets	Cost	Time	Profit
20	0	$ 6,000	100	$ 8,000
0	10	$ 4,000	200	$ 10,000

Alternatively we can build only cabinets, but the 200 weekly hours constrains us to just 10 cabinets per week for a profit of $10,000 (which is better than building 20 tables). In this case we only use $4,000 of our weekly $6,000 budget. So the question becomes if there is some optimal combination of tables and chairs that exhausts both **constraints** and optimizes profits.

A spreadsheet shown below can be used to simply plug and chug values for both **variables** (tables and chairs), making sure both constraints are not violated. Currently building 8 of each item generates the maximum profit of $11,200 and has used up all 200 hours; but 10 tables and 7 cabinets is close at $11,000 and has not exhausted the budget or time yet.

Tables	Cabinets	Cost	Time	Profit
20	0	$ 6,000	100	$ 8,000
19	0	$ 5,700	95	$ 7,600
18	1	$ 5,800	110	$ 8,200
17	2	$ 5,900	125	$ 8,800
16	3	$ 6,000	140	$ 9,400
15	3	$ 5,700	135	$ 9,000
14	4	$ 5,800	150	$ 9,600
13	5	$ 5,900	165	$ 10,200
12	6	$ 6,000	180	$ 10,800
11	6	$ 5,700	175	$ 10,400
10	**7**	$ 5,800	190	**$ 11,000**
9	7	$ 5,500	185	$ 10,600
8	**8**	$ 5,600	200	**$ 11,200**
7	8	$ 5,300	195	$ 10,800
6	8	$ 5,000	190	$ 10,400
5	8	$ 4,700	185	$ 10,000
4	9	$ 4,800	200	$ 10,600
3	9	$ 4,500	195	$ 10,200
2	9	$ 4,200	190	$ 9,800
1	9	$ 3,900	185	$ 9,400
0	10	$ 4,000	200	$ 10,000

So now we would like to know if we can improve on our results, recognizing that fractional units are possible on a weekly basis. Building half a table in a week simply means we can complete it the following week. Instead of continuing with the plug and chug we will develop a classic approach to solving problems like this using graphs. The picture helps in our comprehension of the overall situation, and this approach is easily generalizable to much more complicated problems.

First we list the information in a table:

	Table	Cabinet	Constraints
Cost	$300/unit	$400/unit	$6,000/week
Time	5 hrs/unit	20 hrs/unit	200 hrs/week
Profit	$400/unit	$1,000/unit	

Our constraints can be represented as inequalities:

1. $300 \cdot T + 400 \cdot C \le \$6,000$
2. $5 \cdot T + 20 \cdot C \le 200$

Graphing these lines by computing the intercepts gives us the **feasible region** of possible points (Tables, Chairs) that satisfy the constraints. Note the point (5, 8) is inside the feasible region, but (5, 10) is outside since this combination of 5 tables and 10 cabinets takes 225 hours and violates the time constraint.

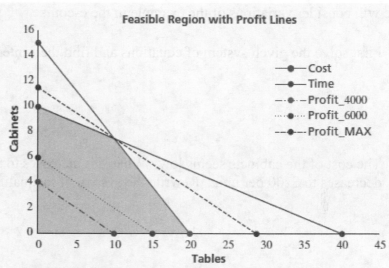

TABLE 13.13 Profit Model Information

$400/unit	$1000/unit	
Table	Cabinet	Profit
10	0	$4000
0	4	$4000
15	0	$6000
0	6	$6000

FIGURE 13.18 Profit Model with Viable Options Satisfying Constraints Shown as Feasible Region

So now we want to maximize our profit function which is linear:

$$P = \$400 \cdot T + \$1,000 \cdot C$$

We can choose any point from this **feasible region** to plug into our profit function. The question is which one gives the most profit. Table 13.13 shows the end points of two possible profit lines: $P = \$4,000$ and $P = \$6,000$. Connecting a pair of endpoints gives a line of possible feasible points with a constant profit. Here are the keys to solving our problem:

- Notice the $6,000 profit line is above the $4,000 profit line.
- Notice that the profit lines are all parallel to each other.
- So profit increases by sliding these lines up.
- We can maximize profit by choosing the last possible profit line touching the feasible region.
- This maximum profit occurs at the corner point where the cost and time lines intersect.

The video in the e-course demonstrates how to find the intersection point giving the maximum profit.

As you read, jot down notes and questions. At the end of this section's Guided Worksheets there is a space for
Reflection *and* ***Monitoring Your Understanding****, where you can try and answer your questions after completing the guided activities and homework.*

Notes & Questions

Guided Practice Activity #1 – Linear Programming

We will consider variations of the example in the e-course.

1. First solve the given system of equations and find their intersection point.

2. The cost of the cabinets seems low. What if it increases to $500 per unit and the table cost decreases to $200 per unit? Rewrite the system of inequalities.

3. Find the new maximum profit.

Guided Practice Activity #2 – Storage Space

You need to buy some cabinets for your office and are considering two brands: A and B. A costs $100 per unit, takes up 6 square feet of space, and holds 15 cubic feet; while B costs $300 per unit, takes up 12 square feet of space and holds 36 cubic feet. You only have $2,100 to spend and cannot use more than 96 square feet of floor space. Find the optimal combination of cabinets to maximize your storage.

1. Find the system of inequalities representing the constraints.

2. Determine which of the following points (A, B) are feasible and find the associated storage.
   (10, 5); (7, 4); (12, 2); (6, 6)

3. Graph the feasible region and find the intersection point of the constraints.

4. Find the maximum storage space.

<table>
<tr><td colspan="2">**Objective 2** – Use Solver</td></tr>
</table>

Key Terms

Solver	
Data analysis	

Summary

Excel has an Add-in package called **Solver** that will automatically solve **linear programming** problems like the one we just solved in the last section. Watch the video for instructions on how to load the Add-in. It will appear in the *Data* menu underneath the **Data Analysis** Add-in that we used in Chapter 12.

To use Solver we first enter the **variables**, constraints, and profit function we want to optimize as shown in the following tables.

TABLE 13.14 Profit Model Variables, Constraints, and Optimization Function

C2		× ✓ *fx*	=A2*300+B2*400	
	A	B	C	D	E
1	Tables	Cabinets	Cost	Time	Profit
2	10	5	$ 5,000	150	$ 9,000

TABLE 13.15 Profit Model Information

	Table	Cabinet	Constraints
Cost	$300/unit	$400/unit	$6000/week
Time	5 hrs/unit	20 hrs/unit	200 hrs/week
Profit	$400/unit	$1000/unit	

Clicking on **Solver** opens the **Parameters** dialog box shown in Figure 13.21. We include the screen shot of the spreadsheet with the **variables, constraints** and profit function in Figure 13.22.

1. First we Set Objective by clicking on cell **E2** which contains the profit function.
2. Select Max so we find the maximum profit.
3. Add constraints by clicking the Add button which opens the Add **Constraint** box shown in Figure 13.23. Enter the appropriate Cell Reference, inequality, and constraint. Click OK.
4. Choose the Simplex LP method and click the Solve button.

That's it! The correct solution will appear in cells **C2:E2**. Solver is a black box in the sense that we don't know how it is finding the solution, thus it is best to familiarize yourself with the solution techniques in the previous objective.

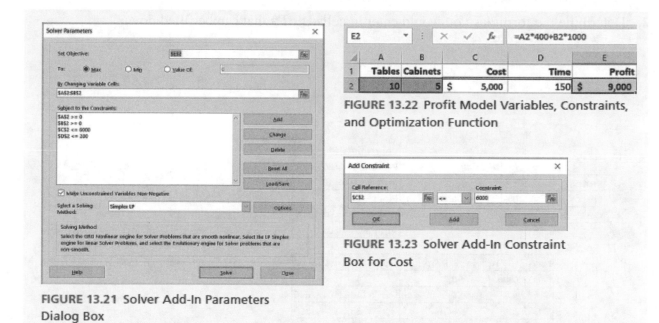

FIGURE 13.22 Profit Model Variables, Constraints, and Optimization Function

FIGURE 13.23 Solver Add-In Constraint Box for Cost

FIGURE 13.21 Solver Add-In Parameters Dialog Box

As you read, jot down notes and questions. At the end of this section's Guided Worksheets there is a space for **Reflection** *and* **Monitoring Your Understanding***, where you can try and answer your questions after completing the guided activities and homework.*

Notes & Questions

Guided Practice Activity #1 – Excel Preview 13.7

Given that Solver requires us to use Excel we will combine the guided activity and Excel Preview.

You feed your animals a mixture of hay and alfalfa. The hay is 5% protein, 40% fiber, and 4 cents per unit. The alfalfa is 20% protein, 10% fiber and 12 cents per unit. For each serving the animals need at least 10 units of protein but not more than 30 units of protein, and at least 20 units of fiber. We want to minimize the cost per serving subject to the constraints.

1. Let h = amount of hay fed and a = amount of alfalfa fed. Write down the system of inequalities for the constraints.

2. Determine the formulas you would enter in cells **D9:F9** that can be filled down.

▲	A	B	C	D	E	F
1					Constraints	
2		Hay	Alfalfa		Protein	<= 30
3	Protein	5%	20%		Protein	>= 10
4	Fiber	40%	10%		Fiber	>= 20
5	Cost	4	12		Hay	>= 0
6					Alfalfa	>= 0
7				a.)		
8		Hay	Alfalfa	Protein	Fiber	Cost
9	1	100	50	15	45	1000
10	2	200	150	40	110	2600
11	3	200	25	15	85	1100
12	4	10	50	10.5	14	640
13	5	75	20	7.75	34	540
14	6	80	90	22	50	1400

3. Which of the rows (1:6) of hay/alfalfa combinations violate the constraints?

4. Sketch the feasible region and also the cost line for Cost = 960 cents.

5. Find the intersection of the lines giving the optimal point.

Section 13.4 Reflections
Monitor Your Understanding

Can you describe what the feasible region represents?

Section 13.5 Statistical (Multivariate) Models

Objective 1 – Explore Regression

Key Terms

Linear regression model	
Multivariate regression model	

Summary

We close this chapter with a brief introduction to multivariate statistical modeling. Chapter 8 covered the basics of regression, looking at the **correlation** between two variables (bivariate).

You are a baseball fan and wonder how a Major League player's batting average is correlated with the number of times they strikeout. We create a scatterplot of the 2017 data, display an equation of the best-fit line and the R^2 value.

The 2017 baseball data is available in *Data Sets*. We can see the possible variables we have to consider when thinking about what contributes to a player's batting average, which is the percentage of times he gets a hit when he bats, and that Jose Altuve led the major leagues with a batting average of 0.346 for the Houston Astros. So got a hit 34.6% of the time he went to bat.

	A	B	C	D	E	F	G	H	I	J	K	L	M	N	O	P	Q	R	S	T	U
1	RK	Player	Team	Pos	G	AB	R	H	2B	3B	HR	RBI	BB	SO	SB	CS	AVG▼	OBP	SLG	OPS	Avg
2	1	Altuve, J	HOU	2B	153	590	112	204	39	4	24	81	58	84	32	6	0.346	0.41	0.547	0.957	346
3	2	Blackmon,	COL	CF	159	644	137	213	35	14	37	104	65	135	14	10	0.331	0.399	0.601	1	331
4	3	Garcia, A	CWS	RF	136	518	75	171	27	5	18	80	33	111	5	3	0.33	0.38	0.506	0.885	330
5	4	Murphy, D	WSH	2B	144	534	94	172	43	3	23	93	59	77	2	0	0.322	0.384	0.543	0.928	322

We can quickly create the scatterplot:

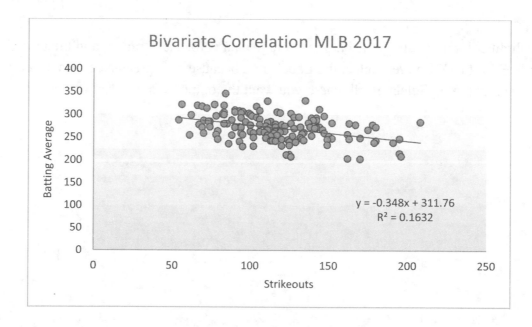

This simple bivariate **correlation** shows us batting averages go down as the number of strikeouts goes up. We have three number to interpret for this **linear regression model**:

1. **Slope** = -0.348
    1. The batting average is decreasing by 0.348 points per strikeout, or about 1 point per 10 strikeouts.
2. **y-intercept** = 311.76
    1. With 0 strikeouts the model predicts a batting average of 312.
3. $R^2 = 0.1632$
    1. 16.3% of the variability in batting average can be explained by variability in strikeouts.

We are now going to create a **multivariate regression model**, meaning we will test multiple factors to see what impact they collectively have on batting average. We can see the possible **variables** we have in Table 13.17. We are going to use the **Data Analysis** Add-in to compute the coefficients of the following multivariate linear equation:

$$Avg = b_0 + b_1 \cdot 2B + b_2 \cdot HR + b_3 \cdot BB + b_4 \cdot SO + b_5 \cdot SB$$

This is called "linear" because none of the variables are squared or cubed etc. Opening Data Analysis we choose Regression.

TABLE 13.17 Multivariate Regression Model Data

	A	B	C	D	E	F
1	Avg	2B	HR	BB	SO	SB
2	346	39	24	58	84	32
3	331	35	37	65	135	14
4	330	27	18	33	111	5
5	322	43	23	52	77	2

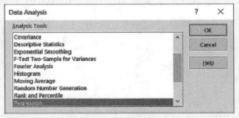

FIGURE 13.27 Data Analysis Add-In Regression

Next highlight column **A** where the batting average data is for the Y range, and then highlight columns **B:F** for the X Range. Select the Labels box because our spreadsheet has headers for the names of each column. Select a cell where you want the output to appear and click OK.

Recall that our **model** is given by the **linear equation**:

$$\text{Avg} = b_0 + b_1 \cdot 2B + b_2 \cdot HR + b_3 \cdot BB + b_4 \cdot SO + b_5 \cdot SB$$

The last part of the output gives us the coefficients and their significance:

	Coefficients	Standard Error	t Stat	P-value	Lower 95%	Upper 95%
Intercept	257.555	12.296	20.947	0.000	233.242	281.867
2B	1.203	0.267	4.501	0.000	0.675	1.732
HR	0.698	0.262	2.669	0.009	0.181	1.215
BB	-0.005	0.103	-0.052	0.959	-0.209	0.198
SO	-0.366	0.072	-5.073	0.000	-0.509	-0.224
SB	0.608	0.205	2.958	0.004	0.201	1.014

$$\text{Avg} = 257.555 + 1.203 \cdot 2B + 0.698 \cdot HR - 0.005 \cdot BB - 0.366 \cdot SO + 0.608 \cdot SB$$

The t Stat is testing the null hypothesis that each coefficient is zero, any *p-value* less than 0.05 allows us to reject this and conclude that coefficient is significant at 0.05. We can see that the walks (BB) are the only coefficient not **significant**, and so can throw that one out. Rerunning the regression without BB slightly changes the coefficients:

$$\text{Avg} = \mathbf{257.443} + \mathbf{1.203} \cdot 2B + \mathbf{0.693} \cdot HR - \mathbf{0.367} \cdot SO + \mathbf{0.607} \cdot SB$$

As you read, jot down notes and questions. At the end of this section's Guided Worksheets there is a space for **Reflection** *and* **Monitoring Your Understanding**, *where you can try and answer your questions after completing the guided activities and homework.*

Notes & Questions

Guided Practice Activity #1 – Country Data

The data for this Excel activity is found in *Data Sets*.

1. For the following scatterplot, interpret the slope, y-intercept, and R-squared value.

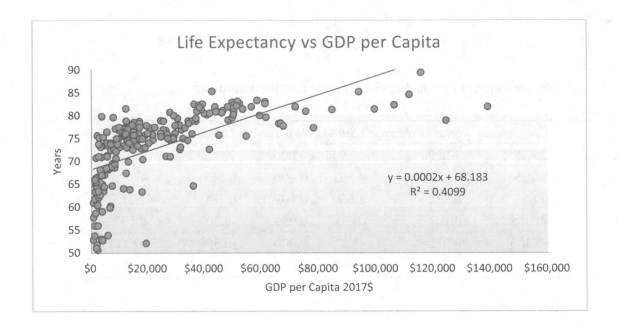

2. Now consider a multivariate model comparing life expectancy to GDP per capita, infant mortality per 1,000 live births, and population growth rates (all 2017 data from CIA factbook).

▲	A	B	C	D	E
1	Country	Life Expectancy	GDP per capita ($1000's)	IM per 1000 live births	Population Growth
2	Afghanistan	51.7	2.00	110.6	2.36
3	Albania	78.5	12.50	11.9	0.31
4	Algeria	77	15.20	19.6	1.7
5	American Samo	73.4	11.20	11.3	-1.3
6	Andorra	82.9	49.90	3.6	0.03

Which gives the following Regression Statistics. Interpret the R-squared value.

Regression Statistics	
Multiple R	0.934056
R Square	0.872461
Adjusted R Square	0.87069
Standard Error	2.872795
Observations	220

3. The following output gives the coefficients of the linear model and their significance. Which of the variables is not significant and should be thrown out?

	Coefficients	Standard Error	t Stat	P-value	Lower 95%	Upper 95%
Intercept	78.182	0.449	174.12	3.2E-234	77.297	79.067
GDP per capita ($1000's)	0.061	0.010	6.45	7.27E-10	0.043	0.080
IM per 1000 live births	-0.311	0.014	-21.56	9.55E-56	-0.340	-0.283
Population Growth	0.067	0.250	0.27	0.788	-0.425	0.560

4. Write down the equation for the multivariate linear model (not using the insignificant statistic), and interpret each coefficient.

<u>**Section 13.5 Reflections**</u>
Monitor Your Understanding

Are you getting more comfortable interpreting the slopes of linear models? Why or why not?

Chapter 1 – An Exel-lent Approach to Relationships

Objective 1.1 – Number Sense
Objective 1.2 – Decimals and Percentages
Objective 1.3 – Formulas and Order of Operations

Objective 1.1 – Number Sense

Key Terms

Quantitative literacy	
Numeracy	
Number	
Numeral	
Number system	
Number line	
Multiplication	
Multiple	
Factor	
Division	

Summary

Quantitative literacy, or **numeracy**, refers to the ability to read, write, and communicate effectively with numbers. We will use number lines to illustrate our numbers. Each line will have units to represent the quantity. Numbers represent the amount of something, how many items or how much of a quantity. Arithmetic refers to operations on numbers, with the following key properties of addition/subtraction and multiplication/division.

Properties of Addition/Subtraction	Properties of Multiplication/Division
1. When adding move to the right on the number line.	1. Multiplication is repeated addition, 4×9 means $9 + 9 + 9 + 9$.
2. When subtracting move to the left on the number line.	2. Multiplication by 1 gives the number, 1×9 means 9.
3. Adding and subtracting the same number results in no change. $3 + 2 - 2 = 3$	3. Multiplication by 0 gives 0, 0×9 means 0 (no nines to add).
4. Adding and subtracting zero also results in no change. $3 + 0 = 3$	4. Multiplication distributes across addition/subtraction, $4 \times (9 - 5) = 4 \times 9 - 4 \times 5$.
5. Adding a negative number is equivalent to subtracting a positive number. $3 - 2 = 3 + -2$	5. Division is the inverse of multiplication, $24 \div 3 = 8$ because $8 \times 3 = 24$.
6. Subtracting a negative number is equivalent to adding that number. $3 - -2 = 3 + 2$	6. Dividing a number by itself gives 1, $24 \div 24 = 1$ because $1 \times 24 = 24$.
	7. Dividing by 0 does not exist (DNE), $24 \div 0 = "?"$ because $? \times 0 \neq 24$.
	8. Taking "one-nth" of a number is the same as dividing by n, taking $\frac{1}{3}$ of $24 = \frac{24}{3}$.

*As you read, jot down notes and questions. At the end of this Guided Worksheet there is a space for **Reflection** and **Monitoring Your Understanding**, where you can try and answer your questions after completing the guided activities and homework.*

Notes & Questions

PKA Practice Problems – Number Lines

Each box of soda contains 4 bottles with 16 ounces per bottle. You have glasses that hold 8 ounces.

1. Draw four number lines, one for boxes, bottles, ounces, and glasses. Represent 3 boxes on each line.

2. Identify half a box on each line.

3. Assume from your 3 boxes you drink 2 bottles. How many boxes/bottles/ounces are left? Indicate the arithmetic operation on each number line.

4. How many bottles are in 7 boxes? How many ounces? Draw number lines indicating these arithmetic operations.

5. If you have 6 boxes of soda and have 8 friends coming over. How many bottles can each friend have? Draw number lines for boxes, bottles and friends indicating this arithmetic operation.

PKA Objective 1.1 Reflections
Monitor Your Understanding

What was something you learned in this section that will make you think differently about your understanding of numbers?

Summary

Understanding our base 10 place value number system is crucial to developing good number sense. Multiplying and dividing by 10 is special in our number system and leads to decimals and percentages. The number 23.74 is represented in the place value tables below. Each number represents a power of 10 and note that we never need more than 9 of any place value since we can trade 10 pennies for a dime etc.

Numerals				2	3	7	4		
Place Values	...	10^4	10^3	10^2	10^1	10^0	10^{-1}	10^{-2}	10^{-3}	...
Place Values	...	10,000	1,000	100	10	1	$\frac{1}{10}$	$\frac{1}{100}$	$\frac{1}{1,000}$	...

We can use monetary values, $23.74, to aid in our understanding.

Numerals		2	3	7	4
Place Values	...	10^2	10^1	10^0	10^{-1}	10^{-2}

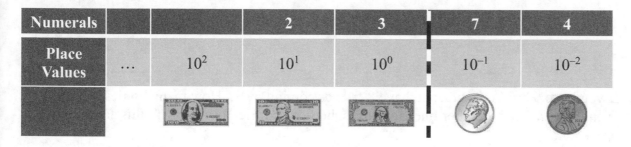

*As you read, jot down notes and questions. At the end of this Guided Worksheet there is a space for **Reflection** and **Monitoring Your Understanding** where you can try and answer your questions after completing the guided activities and homework.*

Notes & Questions

PKA Practice Problems – Base 10 Place Value System

1. Represent $356.19 in a place value table.

Numerals		$\square$	$\square$	$\square$	$\square$	$\square$
Place Values	…	10^2	10^1	10^0	10^{-1}	10^{-2}
		🪙	🪙	🪙	🪙	🪙

2. Represent the arithmetic, $356.19 + $18.56, in a place value table.

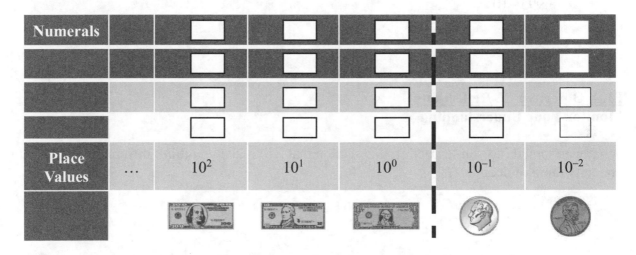

Numerals		$\square$	$\square$	$\square$	$\square$	$\square$
		$\square$	$\square$	$\square$	$\square$	$\square$
		$\square$	$\square$	$\square$	$\square$	$\square$
		$\square$	$\square$	$\square$	$\square$	$\square$
Place Values	…	10^2	10^1	10^0	10^{-1}	10^{-2}
		🪙	🪙	🪙	🪙	🪙

3. Represent the arithmetic, $356.19 – $18.56, in a place value table.

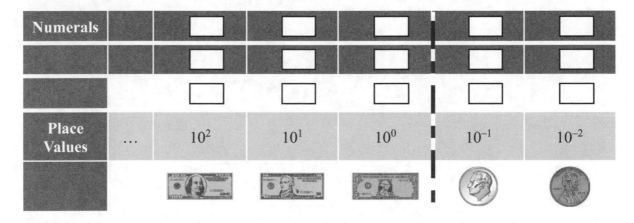

Numerals		$\square$	$\square$	$\square$	$\square$	$\square$
		$\square$	$\square$	$\square$	$\square$	$\square$
		$\square$	$\square$	$\square$	$\square$	$\square$
Place Values	…	10^2	10^1	10^0	10^{-1}	10^{-2}
		🪙	🪙	🪙	🪙	🪙

4. Compute the following:

a. $83.5 \times 100 =$

b. $83.5 \div 1,000 =$

c. $2.71 \times 10^6 =$

d. $473.995 \div 10^2 =$

PKA Objective 1.2 Reflections
Monitor Your Understanding

What was something you learned in this section that will make you think differently about your understanding of decimals?

Objective 1.3 – Formulas and Order of Operations

Summary

The order in which we do things is important. Making a bowl of cereal, brushing your teeth and getting dressed for school requires you do things in proper order. You need to get the bowl in place before pouring the cereal. Similarly you don't want to brush your teeth before you actually put toothpaste on the brush, some things come first! Also groups of things have orders, like you don't want to brush your teeth before eating the cereal, all of the cereal activities come first.

In performing arithmetic it is also important to do things in proper order. Adding 2 children to 3 groups of 5 children (17 children total) is different than adding 2 groups and 3 groups of 5 children each (25 children total). Both of these situations can be represented by: $2 + 3 * 5$, but it is ambiguous as to whether you add first or multiply first. So we have agreed upon an **order of operations**: 1st do everything in parentheses, then take care of exponents, next comes multiplication/division (work from left to right), lastly do addition/subtraction (work from left to right).

*As you read, jot down notes and questions. At the end of this Guided Worksheet there is a space for **Reflection** and **Monitoring Your Understanding**, where you can try and answer your questions after completing the guided activities and homework.*

Notes & Questions

PKA Practice Problems – Order of Operations

Evaluate the following:

- 4+3*6

  [ ]

- (4+3)*6

  [ ]

- 15/5*2

  [ ]

- 15-3+2

  [ ]

- 7+3^2

  [ ]

- (7+3)^2

  [ ]

- 15+9/3+5

  [ ]

- (15+9)/(3+5)

  [ ]

The spreadsheet has 4 input cells: **E2:E5**.

Compute the weighted average for the new GPA estimate to 2 decimal places:

$$\frac{(\text{Credits so Far}) \cdot \text{GPA}_{now} + (\text{New Credits}) \cdot \text{GPA}_{semester}}{(\text{Credits so Far}) + (\text{New Credits})}$$

	A	B	C	D	E
1					
2			Credits so Far:		72
3			GPA so Far:		2.54
4			Credits this Semester:		16
5			Estimated GPA for this Semester:		3.0
6					
7		a.)	New GPA estimate:		2.62

1. Enter the formula you would type into Excel in cell E7 for the new GPA estimate. Try and use as few parentheses as possible.

2. What estimated GPA for the semester in cell E5 will give a new GPA estimate in cell E7 of 2.75? How would Excel make this easier to find?

PKA Objective 1.3 Reflections
Monitor Your Understanding

What was something you learned in this section that will make you think differently about your understanding of order of operations?

Chapter 2 – Ratios and Proportions
Prior Knowledge Activation (PKA)

Objective 2.1 – Number Words
Objective 2.1 – Number Words
Objective 2.2 – Fractions
Objective 2.3 – Setting Up Proportions
Objective 2.4 – Basic Equation Solving

Objective 2.1 – Number Words

Summary

A **number** is the amount of something, how many items or how much of a quantity. Data tables frequently will report numbers to be in "thousands" or "millions" or "billions" in order to save space and not have to write out so many zeroes. It is important you pay attention to such number words as missing them or simply not knowing how to interpret them can lead to gross misinterpretations.

If we say 2 thousand it clearly means 2,000 but people get confused with 2,325 thousand. The word thousand tells us to multiply by 1,000 or simply add three zeroes: 2,325 thousand = 2,325,000 which is 2 million 325 thousand, or 2,000,000 + 325,000.

Number Words	Zeroes				Excel Notation
Thousands	3	2,325 thousand	2,325,000	$2,325 \times 10^3$	2,325E+3
Millions	6	2,325 million	2,325,000,000	$2,325 \times 10^6$	2,325E+6
Billions	9	2,325 billion	2,325,000,000,000	$2,325 \times 10^9$	2,325E+9
Trillions	12	2,325 trillion	2,325,000,000,000,000	$2,325 \times 10^{12}$	2,325E+12

Notice how using powers of ten in the table also conserves space and is a common practice, especially in scientific journals. Thus it is important for you to realize that multiplying by powers of 10 simply adds zeroes. When doing arithmetic we also use the powers of 10. To multiply, we add the powers of 10:

$$8 \text{ billion} \times 2 \text{ thousand} = (8 \times 10^9) \times (2 \times 10^3) = 16 \times 10^{12} = 16 \text{ trillion}$$

To divide, we subtract the powers of 10:

$$8 \text{ billion} \div 2 \text{ thousand} = (8 \times 10^9) \div (2 \times 10^3) = 4 \times 10^6 = 4 \text{ million}$$

*As you read, jot down notes and questions. At the end of this Guided Worksheet there is a space for **Reflection** and **Monitoring Your Understanding**, where you can try and answer your questions after completing the guided activities and homework.*

<u>**Notes & Questions**</u>

<u>**PKA Practice Problems #1**</u>– Wordy Numbers

Perform the following arithmetic and represent answers in number words (1 decimal place). Note that there are multiple possible answers so please keep the numbers between 1 and 10,000 and use the smallest number word possible (or no word). Some examples may help:

$48.3 \text{ thousand} \times 5250 \text{ thousand} = 25{,}375 \times 10^6 = 25.4 \times 10^9 = 25.4 \text{ billion}$

$48.3 \text{ billion} \div 5250 \text{ thousand} = 0.0092 \times 10^6 = 9.2 \times 10^3 = 9.2 \text{ thousand}$

$48.3 \text{ billion} + 5250 \text{ thousand} = 48{,}305{,}250{,}000 = 48.3 \text{ billion}$

1. $125 \text{ million} \times 2{,}300 \text{ thousand}$

2. $125 \text{ million} + 2{,}300 \text{ thousand}$

3. $125 \text{ million} \div 2{,}300 \text{ thousand}$

4. $2.4 \text{ billion} \div 450 \text{ thousand}$

5. $2.4 \text{ billion} \times 450 \text{ thousand}$

6. $2.4 \text{ billion} - 450 \text{ thousand}$

7. $510 \text{ trillion} \div 4.2 \text{ million}$

PKA Practice Problems #2– Wordy Data

We saw the following data in the text.

Year	Motor Vehicle Deaths (MVD)	Population (Thousands)	Licensed Drivers (Thousands)	Registered Motor Vehicles (Thousands)	Vehicle Miles Traveled (Billions)
2002	43,005	287,625	194,602	225,685	2,856
2011	32,367	311,592	211,875	257,512	2,946
2015	35,092	321,419	218,084	281,382	3,095

1. How many more drivers were there in 2015 than 2002?

2. Does it make sense to subtract the number of drivers in 2015 from the number of vehicles in 2015? Why or why not?

3. Divide the number of vehicles by the population in 2015 and 2002 (3 decimal places). Which answer is bigger?

4. What do the answers in #3 mean?

5. Divide the number of miles traveled by number of vehicles in 2015 and 2002 (3 decimal places)? Which is bigger?

6. What do the answers in #5 mean?

7. Does it makes sense to add the number of motor vehicle deaths from year to year? If yes what does the sum mean, if no why not?

8. Does it makes sense to add the populations from year to year? If yes, what does the sum mean; or if no, why not?

9. Does it make sense to subtract the number of motor vehicle deaths from year to year? To subtract the populations?

PKA Objective 2.1 Reflections
Monitor Your Understanding

What was something you learned in this section that will make you think differently about your understanding of numbers?

Objective 2.2 – Fractions

<u>**Key Terms**</u>

Fraction	
Unit fraction	
Equivalent fractions	

<u>**Summary**</u>

The word fraction comes from "fractus" to break, while numerator comes from "numerare", to number, and denominator from "nomen", to name. Thus the denominator represents the name of the parts (halves, thirds, fourths, fifths, etc.) while the numerator simply counts out these parts. We saw in chapter 1 how to break a class into 3 groups or 2 teams:

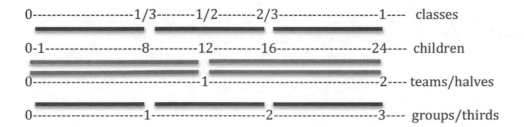

```
0--------------------1/3--------1/2--------2/3--------------------1----  classes

0-1--------------------8----------12----------16--------------------24----  children

0--------------------------------1----------------------------------2----  teams/halves

0--------------------1----------------------2--------------------3----  groups/thirds
```

CAUTION! Note that halves and thirds are units just like groups and teams!

Two groups of 8 children can thus be thought of as the number $\frac{2}{3}$ of a class of 24 children, or the number 2 thirds. It is important for you to understand that both $\frac{2}{3}$ class and 2 thirds are equivalent to 16 children:

$$\frac{2}{3} \text{ class} \equiv 2 \text{ thirds} \equiv 16 \text{ children}$$

Fractions are **equivalent** if we can multiply the numerator and denominator of one fraction by the same factor to get the other fraction. We use an equal sign to denote the equality of the two numbers.

$$\frac{a}{b} = \frac{a \cdot \boldsymbol{n}}{b \cdot \boldsymbol{n}}$$

This definition explains why we can factor out like terms to simplify fractions, the common factor cancels or "divides away". Let's subtract 3 fourths – 7 twentieths and reduce the answer to simplest terms. Just as we can't subtract cats and dogs, we cannot subtract fourths and twentieths. We need to rewrite 3 fourths with its equivalent fraction 15 twentieths:

$$\frac{3}{4} = \frac{3 \cdot \boldsymbol{5}}{4 \cdot \boldsymbol{5}} = \frac{15}{20}$$

3 fourths – 7 twentieths = 15 twentieths – 7 twentieths   =   8 twentieths.
Note that when we write this sum using fractional notation we only add/subtract the numerators:

$$\frac{3}{4} - \frac{7}{20} = \frac{15}{20} - \frac{7}{20} = \frac{8}{20}$$

Remember that the denominator is the "name" or unit. We can reduce the final answer to $\frac{2}{5}$ by recognizing that both 8 and 20 are multiples of 4 and using the notion of equivalent fractions:

$$\frac{8}{20} = \frac{2 \times \boldsymbol{4}}{5 \times \boldsymbol{4}} = \frac{2}{5}$$

*As you read, jot down notes and questions. At the end of this Guided Worksheet there is a space for **Reflection** and **Monitoring Your Understanding**, where you can try and answer your questions after completing the guided activities and homework.*

Notes & Questions

PKA Practice Problems #1– Cheery-O!

A box of Cheerios cereal has 72 ounces or 9 cups. A serving size is 1 cup which contains 100 calories, 3 g of protein, 140 mg of sodium.

1. Fill in the numbers for the lines for boxes, ninths, cups, ounces, calories, grams of protein and mg of sodium with 9 tick marks on each line for serving sizes.

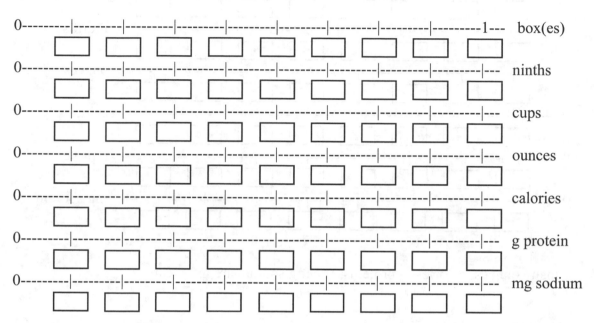

2. Fill in the numbers for the line for thirds of a box. And compute the number of cups, ounces, calories, protein, and sodium in one-third of a box.

3. How many cups and mg of sodium are in 2 ninths plus 1 third of a box?

4. You have 3 fourths of a box and eat 1 third of a box. How much of the box is left?

5. You give your toddler a quarter of a cup of Cheerios. What fraction of a box is this and how many ounces, calories, g protein and mg sodium do they eat?

6. A cup has 8 ounces. Fill in the numbers for the lines for cups, eighths, fourths, ounces, calories, grams of protein and mg of sodium with 8 tick marks on each line. Represent answers as fractions in lowest possible terms.

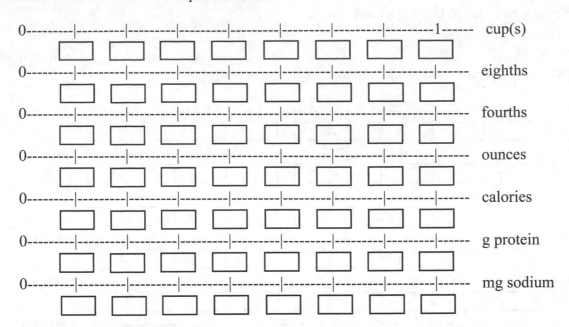

7. How many ounces, calories, g proteins, and mg sodium are in $\frac{3}{4}$ cup?

PKA Objective 2.2 Reflections
Monitor Your Understanding

What was something you learned in this section that will make you think differently about your understanding of fractions?

Objective 2.3 – Setting up Proportions

Summary

We have seen that we can only add and subtract numbers with the same units, and this means adding and subtracting fractions only with the same denominator. We used the notion of equivalent fractions to rewrite 3 fourths as 15 twentieths:

$$\frac{3}{4} = \frac{3 \cdot 5}{4 \cdot 5} = \frac{15}{20}$$

When two fractions are equal we can cross multiply numerators and denominators and **know** these products are equal:

$$\frac{3}{4} = \frac{15}{20} \ldots \qquad \frac{3 \times 20}{4 \times 20} = \frac{15 \times 4}{20 \times 4} \ldots \qquad 3 \times 20 = 15 \times 4$$

This turns out to be an important solution technique when working with ratios. Setting two ratios equal to each other is called setting up a proportion.

Given the ratio of boys to girls in a class is three to four, 3 : 4, and there are 21 boys, we can determine how many girls are in the class by setting up a proportion.
We can set up the proportion three is to four as 21 is to x:

$$\frac{3}{4} = \frac{21}{x}$$

And cross multiply to get an equation to solve for the unknown numerator:

$$3 \cdot x = 4 \cdot 21$$

To solve this equation for the unknown variable x, we need to get it by itself on one side of the equation. So we need to undo the multiplication by dividing both sides by 3:

$$x = \frac{4 \cdot 21}{3} = \frac{4 \cdot 7 \cdot 3}{3} = 28$$

Notice that we could have seen in the proportion that 21 is 3 times 7, and thus x must be 4×7:

$$\frac{3}{4} = \frac{21}{x} = \frac{3 \cdot 7}{4 \cdot 7}$$

*As you read, jot down notes and questions. At the end of this Guided Worksheet there is a space for **Reflection** and **Monitoring Your Understanding**, where you can try and answer your questions after completing the guided activities and homework.*

Notes & Questions

The ratio of the weight of grapes to the cost is 3 lbs : $5.

1. If you buy 12 lbs how much does it cost?

2. If you buy 2.5 lbs how much does it cost?

3. How many pounds can you buy with $9?

4. How many pounds can you buy with $15?

5. How much do grapes cost per pound? Ratios represented as "something *per* something" are called rates and will be studied in chapter 3.

6. How many pounds can you buy per dollar?

7. Another store has 2 lbs : $4, is this a better buy? Why?

PKA Practice Problems #2– Dependency

The Census Bureau computes what are called "dependency" ratios, comparing the size of working age people (18 – 64) to dependent age people: elderly (65 +) and children (< 18).

1. The number (in thousands) of working age people is given in the table. If the ratio of working to elderly in 2011 is 9 : 2, compute the number of elderly. Round to nearest thousand. Hint: Set up a proportion!

	2011	2050
18 - 64	196,264	230,147
65 +		

2. The number of elderly is expected to be 83,739,000 in 2050. Set up a proportion to compute the dependency ratio in 2050, scaling the second quantity in the ratio to 2 (round final answer to 1 decimal place).

3. Use the ratio from #2 in a sentence.

4. How do these ratios indicate the "aging" of America?

5. The October 2017 *Drug Threat Assessment Report* from the Drug Enforcement Agency (DEA), part of the Department of Justice, gives the following figure showing the striking rise of drug over dose deaths in the US:

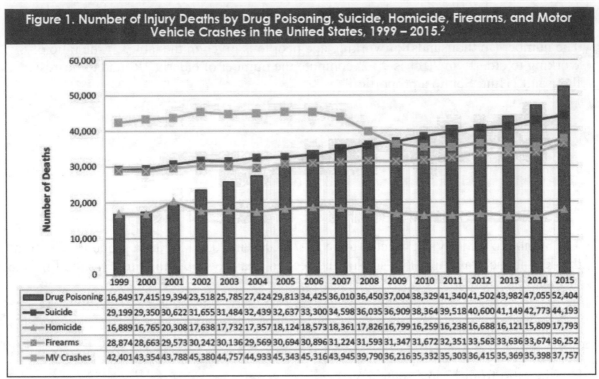

Figure 1. Number of Injury Deaths by Drug Poisoning, Suicide, Homicide, Firearms, and Motor Vehicle Crashes in the United States, 1999 – 2015.[2]

	1999	2000	2001	2002	2003	2004	2005	2006	2007	2008	2009	2010	2011	2012	2013	2014	2015
Drug Poisoning	16,849	17,415	19,394	23,518	25,785	27,424	29,813	34,425	36,010	36,450	37,004	38,329	41,340	41,502	43,982	47,055	52,404
Suicide	29,199	29,350	30,622	31,655	31,484	32,439	32,637	33,300	34,598	36,035	36,909	38,364	39,518	40,600	41,149	42,773	44,193
Homicide	16,889	16,765	20,308	17,638	17,732	17,357	18,124	18,573	18,361	17,826	16,799	16,259	16,238	16,688	16,121	15,809	17,793
Firearms	28,874	28,663	29,573	30,242	30,136	29,569	30,694	30,896	31,224	31,593	31,347	31,672	32,351	33,563	33,636	33,674	36,252
MV Crashes	42,401	43,354	43,788	45,380	44,757	44,933	45,343	45,316	43,945	39,790	36,216	35,332	35,303	36,415	35,369	35,398	37,757

Source: Centers for Disease Control Prevention

The Census estimated the population in 2015 (on July 1) to be 321,419,820 people. Compute the ratio of drug overdose deaths to the population with the second quantity scaled to 100,000 (1 decimal place) by firs setting up the appropriate proportion.

6. The Census gave the population in 2000 (on April 1) to be 281,421,906 people. Compute the ratio of drug overdose deaths to the population with the second quantity scaled to 100,000 (1 decimal place).

7. Someone could argue that since the population is increasing, the number of drug overdoses would also be expected to increase. How do the ratios you computed show the problem is getting worse?

8. Compute the number of drug overdose deaths per day in 2015 (1 decimal place).

9. Compute a ratio of your choice that gives evidence to the drug overdose crisis we are experiencing in this country.

10. If the ratio of 2016 drug overdose deaths (DOD) to 2015 is the same as the ratio of 2015 DOD to 2014, what is the estimated number of DOD in 2016 (whole number)?

11. The actual number is 2016 was about 64,000 DOD. How does this compare to your estimate in #10 and what does it mean in terms of the problem getting better or worse?

PKA Objective 2.3 Reflections
Monitor Your Understanding

What was something you learned in this section that will make you think differently about your understanding of ratios?

Summary

In setting up a proportion and cross multiplying we have seen how to solve an equation for an unknown variable:

$$3 \cdot x = 4 \cdot 21 \quad \ldots \quad \ldots \quad x = \frac{4 \cdot 21}{3} = \frac{4 \cdot 7 \cdot 3}{1 \cdot 3} = 28$$

Next we will consider equations with two variables. Assume each brick weighs 3 lb. Determine an equation for the weight of any number of bricks. We can make a table to see the relationship:

Number of bricks	Weight of bricks (lb)
1	3
2	6
3	9

And clearly see that the weight is always 3 times the number:

$$w = 3 \cdot n \quad \ldots \quad \ldots \quad y = 3 \cdot x$$

Students are often used to seeing y as a function of x, but it makes more sense to choose variables (letters) that represent the quantities. If we have 12 bricks we can quickly plug into our equation to find the weight: $w = 3 \cdot 12 = 36$ lb. But if we are given a weight like 24 lb then we have to work backwards to find the number by dividing by 3:

$$24 = 3 \cdot n \quad \ldots \quad \ldots \quad \frac{24}{3} = 8 = n$$

Thus the number of bricks will always be one-third the weight:

$$n = \frac{1}{3} \cdot w$$

Note that 1/3 in the equation:

$$n = \frac{1}{3} \cdot w$$

is the reciprocal of 3 in the equation:

$$w = 3 \cdot n$$

*As you read, jot down notes and questions. At the end of this Guided Worksheet there is a space for **Reflection** and **Monitoring Your Understanding**, where you can try and answer your questions after completing the guided activities and homework.*

Notes & Questions

PKA Practice Problems #1– Grape-ful Take 2

The ratio of the weight of grapes to the cost is 3 lbs : $5.

1. Fill in the following table.

Weight (lbs)	Cost ($)
1	☐
2	☐
3	☐
4	☐
5	☐
6	☐

2. What is the equation for cost as a function of the weight (use a fraction for the number)?

3. Fill in the following table:

Cost ($)	Weight (lbs)
1	☐
2	☐
3	☐
4	☐
5	☐
6	☐

4. What is the equation for weight as a function of cost (use a fraction)?

5. How are the fractions in the two equations related?

6. The time (hours) it takes you to drive to school (7 miles away) is a function of your speed or rate (mph), given by the equation:

$$t = \frac{7}{r}$$

a. If you average 28 mph, how long will it take?

b. If it takes you 12 minutes how fast do you drive?

c. Find the equation for rate as a function of time.

PKA Objective 2.4 Reflections
Monitor Your Understanding

What was something you learned in this section that will make you think differently about your understanding of equations and proportions?

Chapter 3 – Units, Conversions, Scales, and Rates
Prior Knowledge Activation (PKA)

Objective 3.1 – Arithmetic with Units
Objective 3.2 – Fraction Arithmetic
Objective 3.3 – Scientific Notation

Objective 3.1 – Arithmetic with Units

Summary

A **number** is the amount of something, how many items or how much of a quantity, and thus will always have **units**. Recall our key rule that we can only add/subtract numbers with the *same* units. Interestingly we can multiply/divide numbers with *different* units. Assume you go on a 120 mile trip and use 5 gallons of gas. We can determine how many miles you can go with 1 gallon of gas and how many gallons you would use on 4 such trips.

First let's draw in number lines for the different units.

$$0\text{----?--------------------}\textbf{120}\text{------------------------}240\text{---------} \quad \text{miles}$$
$$0\text{----}\textbf{1}\text{----}2\text{----}3\text{----}4\text{----}\textbf{5}\text{----}6\text{----}7\text{----}8\text{----}9\text{----}10\text{----}11\text{----} \quad \text{gallons}$$

It makes no sense to add miles and gallons right? 120 miles + 5 gallons = 125 miles + gallons? But it does make sense to divide them:

$$\frac{120 \text{ miles}}{5 \text{ gallons}} = \frac{\textbf{24} \text{ miles}}{1 \text{ gallon}} = 24\frac{\text{mi}}{\text{gal}} = 24 \text{ miles per gallon (mpg)}$$

Notice how this division is like setting up a **proportion** and solving for 24. The number lines behave like horizontal **ratio tables**. Dividing miles by gallons we get a new **compound unit** miles per gallon (mi/gal or mpg).

If we go on 4 such trips we would add:
$$5 \text{ gal} + 5 \text{ gal} + 5 \text{ gal} + 5 \text{ gal} = 4 \times 5 \text{ gal} = 20 \text{ gal}.$$

Note that multiplication is simply repeated addition. Above we stated that numbers will always have units, but the 4 in the multiplication was written without units. The number 4 represents 4 trips and if we thinks of the 5 as representing 5 gallons per trip we get:

$$4 \text{ \st{trips}} \times 5\frac{\text{gal}}{\text{\st{trip}}} = 20 \text{ gal}$$

The units, *trips*, cancel in the multiplication, behaving essentially like factors.
Physics is notorious for all sorts of strange units. Work is force times distance, so applying 2 pounds of force to push an object 3 feet results in:

$$3 \text{ ft} \times 2 \text{ lbs} = 6 \text{ ft lbs}$$

Speeding up by 2 m/s per second gives an acceleration of 2 m/s^2. Basically we can multiply and divide by any units we wish (canceling units when possible just like common factors), resulting in some strange units!

*As you read, jot down notes and questions. At the end of this Guided Worksheet there is a space for **Reflection** and **Monitoring Your Understanding**, where you can try and answer your questions after completing the guided activities and homework.*

Notes & Questions

PKA Practice Problems #1– Know Your Units

Perform the following arithmetic and indicate the proper units of the answer, or indicate the arithmetic cannot be performed.

1. You buy 3 pounds of grapes for $12.

   a. $\$12 \div 3 \text{ lbs}$

   b. $\$12 - 3 \text{ lbs}$

   c. $\$12 \times 3 \text{ lbs}$

   d. $3 \text{ lbs} \div \$12$

2. You are walking at a rate of 4 feet per second.

   a. $4 \text{ ft/sec} \times 8 \text{ sec}$

   b. $20 \text{ ft} \times 4 \text{ ft/sec}$

c. $20 \text{ ft} \div 4 \text{ ft/sec}$

d. $4 \text{ ft/sec} + 8 \text{ sec}$

e. $4 \text{ ft/sec} \times 5 \text{ min}$

3. A watt (W) is a unit of power, telling you the rate of energy being used in joules (J) per second (1 W = 1 J/sec). A kilowatt (kW) is 1,000 watts. We can convert joules to food calories using 4,186.8 J = 1 Cal, both are units of energy. Use joules for answers.

a. $100 \text{ W} \times 5 \text{ sec}$

b. $1 \text{ kW} \times 1 \text{ hour}$

c. $1 \text{ W} + 5 \text{ kW}$

d. $100 \text{ W} \div 5 \text{ J}$

PKA Objective 3.1 Reflections
Monitor Your Understanding

What was something you learned in this section that will make you think differently about your understanding of units?

Summary

Recall that for **fractions** the denominator can serve as the **unit**, and thus we can only add/subtract fractions with common denominators. It is possible to multiply/divide fractions with different denominators, just as it is possible to multiply/divide numbers with different units.

Assume you have $\dfrac{9}{4}$ cups of laundry detergent and your roommate has $\dfrac{6}{4}$ of a cup. How many cups of detergent do you have combined? If it takes $\dfrac{3}{4}$ cup of detergent to run a load of wash, and you need to do 4 loads of wash do you have enough detergent?

We can add the fractions since they have a common denominator,
9 fourths + 6 fourths = 15 fourths:

$$\frac{9}{4}+\frac{6}{4}=\frac{15}{4}=3\frac{3}{4} \text{ cups}$$

Next we compute the number of cups for 4 loads of wash, 4×3 fourths $=12$ fourths:

$$4 \ \cancel{\text{loads}} \times \frac{3 \text{ cups}}{4 \ \cancel{\text{load}}} = \frac{4\times3}{4} \text{ cups} = \frac{12}{4} \text{ cups}$$

So, we have 15 fourths cups and need 12 fourths. Therefore, yes, we have enough detergent for 4 loads. We have seen how to multiply a fraction by a whole number. How about multiplying a fraction by a fraction? Recall that taking one-third of a quantity is equivalent to multiplying by the unit fraction $\dfrac{1}{3}$ or dividing the quantity by 3.

You have $\dfrac{15}{4}$ cups of detergent and use $\dfrac{2}{3}$ of your detergent. How many cups do you use?

First we take one third of $\dfrac{15}{4}$ by dividing 15 by 3:

$$\frac{1}{3}\times15 \text{ fourths} = \frac{15}{3} \text{ fourths} = 5 \text{ fourths}$$

The notion of **equivalent fractions** leads to a shortcut:

$$\frac{1}{3}\times\frac{15}{4}=\frac{1\times3\times5}{3\times4}=\frac{5}{4}$$

We essentially can multiply "straight across", numerators and denominators, and then factor out the 3. If 1 third is 5 fourths then 2 thirds must be 10 fourths, and once again we see that multiplying straight across works:

$$\frac{2}{3} \times \frac{15}{4} \text{ cups} = \frac{2 \times 15}{3 \times 4} = \frac{2 \times 3 \times 5}{3 \times 4} = \frac{\mathbf{3} \times 2 \times 5}{\mathbf{3} \times 4} = \frac{2 \times 5}{4} = \frac{10}{4} \text{ cups}$$

Therefore, we use 10 fourths, which we can simplify again using equivalent fractions:

$$\frac{10}{4} = \frac{5 \times \mathbf{2}}{2 \times \mathbf{2}} = \frac{5}{2} = 2\frac{1}{2} \text{ cups}$$

Now you may have recognized that 15 was evenly divisible by 3 in the last example. What would happen if we wanted to take $\frac{2}{3}$ of $\frac{5}{4}$? In this case we would first find an equivalent fraction of $\frac{5}{4}$ with a numerator that is a multiple of 3:

$$\frac{2}{3} \times \frac{5}{4} \text{ cups} = \frac{2}{3} \times \frac{5 \times 3}{4 \times 3} = \frac{2 \times 15}{3 \times 4 \times 3} = \frac{2 \times 3 \times 5}{3 \times 4 \times 3} = \frac{\mathbf{3} \times 2 \times 5}{\mathbf{3} \times 4 \times 3} = \frac{2 \times 5}{4 \times 3} = \frac{10}{12} \text{ cups}$$

Note that once again our rule of multiplying straight across works!

Rule: When multiplying fractions, simply multiply straight across:

$$\frac{a}{b} \times \frac{c}{d} = \frac{a \times c}{b \times d}$$

Recall from the definition of division: $m \div n$ can be thought of as taking "one-n^{th}" of the number and can be represented as the **fraction,** $\frac{m}{n}$. This *quotient* can be interpreted in two different ways:

1. Divide m into n equal parts, giving the size of each part.
2. The number of groups of size n can we "take" from m.

So $8 \div 2 = 2$ equal parts of size 4, or 4 groups of size 2. When dividing by a fraction, it only makes sense to use the second interpretation.

Assume you have $\frac{15}{4}$ cups of detergent and need $\frac{3}{4}$ cup for a load of wash. How many loads can you do? We are being asked to find the number of ¾ we can take from $\frac{15}{4}$, so this is a division problem. The denominator acts like **units**, allowing us to simply divide the whole number numerators:

$$15 \text{ fourths of a cup} \div 3 \text{ fourths of a cup} = \frac{15 \text{ fourths}}{3 \text{ fourths}} = 5 \text{ loads}$$

$$\frac{15}{4} \div \frac{3}{4} = 15 \div 3$$

The key to dividing fractions seems to be having a common denominator. If we had $\frac{15}{4}$ cups and each load took $\frac{2}{3}$ cup we would first have to find common denominators:

$$\frac{15}{4} \div \frac{2}{3} = \frac{15 \times 3}{4 \times 3} \div \frac{2 \times 4}{3 \times 4} = 15 \times 3 \div 2 \times 4 = \frac{15 \times 3}{2 \times 4} = \frac{15}{4} \times \frac{3}{2}$$

We simply multiply by the reciprocal!

> **Rule:** Dividing by a fraction is equivalent to multiplying by the reciprocal:
> $$\frac{a}{b} \div \frac{c}{d} = \frac{a}{b} \times \frac{d}{c}$$

*As you read, jot down notes and questions. At the end of this Guided Worksheet there is a space for **Reflection** and **Monitoring Your Understanding**, where you can try and answer your questions after completing the guided activities and homework.*

Notes & Questions

PKA Practice Problems #1– Fraction Fun

Perform the following arithmetic and indicate the fraction equation used.

1. You have $\frac{2}{3}$ gallon of juice and drink half of it. How much is left?

2. You have $\frac{3}{5}$ gallon of juice and drink half of it. How much is left?

3. You have $\frac{7}{5}$ gallons of juice and want to fill cups that hold $\frac{1}{10}$ of a gallon. How many cups can you fill?

4. You have $\frac{16}{5}$ gallons of juice and want to fill pitchers that hold $\frac{2}{3}$ gallon each. How many pitchers can you fill?

5. You have $\frac{3}{4}$ gallon of juice and add $\frac{1}{5}$ gallon to it. How much do you have?

6. You have $\frac{3}{5}$ gallon of juice and drink $\frac{1}{3}$ gallon. How much is left?

PKA Objective 3.2 Reflections
Monitor Your Understanding

What was something you learned in this section that will make you think differently about your understanding of fraction arithmetic?

Key Terms

Scientific notation	

Summary

We have seen how to deal with **number** words like 3,250 thousand = 3,250,000. Tables of data will often report data "in thousands" so they don't have to waste space on the zeroes. When computing with such large numbers it is helpful to use scientific notation. To perform the following arithmetic we use the associative and commutative properties of multiplication:

$$3,250,000 \times 0.000047 = \left(3.25 \times 10^6\right) \cdot \left(4.7 \times 10^{-5}\right)$$
$$= \left(3.25 \times 4.7\right) \cdot \left(10^6 \times 10^{-5}\right)$$
$$= \left(15.275\right) \cdot \left(10^{6-5}\right)$$
$$= 15.275 \times 10^1 = 1.5275 \times 10^2$$

To divide, we again group the powers of 10:

$$3,250,000 \div 0.000047 = \frac{\left(3.25 \times 10^6\right)}{\left(4.7 \times 10^{-5}\right)}$$
$$= \frac{3.25}{4.7} \times \frac{10^6}{10^{-5}}$$
$$= 0.6915 \times 10^{6-(-5)} = 6.915 \times 10^{10}$$

Note that in order to write the answer in scientific notation we needed to both move the decimal place and adjust the power of 10.

CAUTION! Moving the decimal place left increases the power of 10, and moving the decimal place right decreases the power of 10.

*As you read, jot down notes and questions. At the end of this Guided Worksheet there is a space for **Reflection** and **Monitoring Your Understanding**, where you can try and answer your questions after completing the guided activities and homework.*

Notes & Questions

PKA Practice Problems #1– Scientific Notation Drill

Perform the following arithmetic by first rewriting each number in scientific notation, and express the answer in scientific notation (3 decimal places for final answer).

1. $3{,}250{,}000{,}000 \times 0.0047$

2. $3{,}250{,}000{,}000 \times 0.0047$

3. 10.2 trillion ÷ 329 milion

4. 10.2 trillion × 329 milion

5. 29,654 thousand × 417

6. 417 thousand ÷ 29,654

7. 712,345 thousand × 0.0012

8. **1.2** thousand ÷ 712,345

PKA Practice Problems #2– Powers of 10 MVD

The following table from the e-text gives motor vehicle deaths (MVD) as various rates.

TABLE 3.6 U.S. Department of Transportation National Highway Traffic Safety Administration Traffic Safety Facts 2015

Year	Fatalities	Resident Population (Thousands)	Fatality Rate per 100,000 Population	Licensed Drivers (Thousands)	Fatality Rate per 100,000 Licensed Drivers	Registered Motor Vehicles (Thousands)	Fatality Rate per 100,000 Registered Vehicles	Vehicle Miles Traveled (Billions)	Fatality Rate per 100 Million Vehicle Miles Traveled
2002	43,005	287,625	14.95	194,602	22.1	225,685	19.06	2856	1.51
2010	32,999	309,347	10.67	210,115	15.71	257,312	12.82	2967	1.11
2011	32,479	311,719	10.42	211,875	15.33	265,043	12.25	2950	1.1
2015	35,092	321,419	10.92	218,084	16.09	281,312	12.47	3095	1.13

Data from: Published April 2017 Data, https://crashstats.nhtsa.dot.gov/Api/Public/ViewPublication/812384

1. Verify the fatality rate per 100 million vehicle miles traveled in 2002.

   a. First divide the number of MVD in 2002 by the number of miles traveled and represent answer in scientific notation. This gives the number of MVD per mile.

   b. Now multiply numerator and denominator by 100 million to get the rate per 100 million miles.

2. Verify the fatality rate per 100,000 population in 2015.

   a. First divide the number of MVD in 2015 by the number of people and represent answer in scientific notation. This gives the number of MVD per person.

   b. Now multiply numerator and denominator by 100 thousand to get the rate per 100 thousand people.

PKA Objective 3.3 Reflections
Monitor Your Understanding

What was something you learned in this section that will make you think differently about your understanding of using powers of 10 to perform arithmetic?

Chapter 4 – Percentages
Prior Knowledge Activation (PKA)

Objective 4.1 – Percent Proportions

<u>Summary</u>

The notion of **equivalent fractions** allows us to represent any **fraction** in an infinite number of ways:

$$\frac{3}{4} = \frac{3 \cdot n}{4 \cdot n}$$

In particular, we can multiply the denominator by a factor to get 100, and represent our fraction as a **percentage**:

$$\frac{3}{4} = \frac{3 \cdot 25}{4 \cdot 25} = \frac{75}{100} = 75\%$$

We often interpret percentages as **part-to-whole ratios**, allowing us to set up what is called a "percent proportion":

$$\frac{\text{part}}{\text{whole}} = \frac{\text{percent}}{100}$$

CAUTION! We can scale the second quantity in **any** ratio to 100, not just part-to-whole ratios.

Assume you go on a 120 mile trip and after 48 miles make a rest stop. What percentage of the distance have you traveled? First let's set up the percent proportion:

$$\frac{48 \text{ miles}}{120 \text{ miles}} = \frac{\text{percent}}{100}$$

To solve for the **percentage** we use our tried and true technique of cross multiplying.

$$\frac{48 \text{ miles}}{120 \text{ miles}} = \frac{\text{percent}}{100}$$

The units cancel and we get:

$$48 \times 100 = 120 \times \text{percent}$$

$$\text{percent} = \frac{4,800}{120} = 40$$

So we have traveled 40% of the distance after 48 miles. Note the usage of the word "of" in the last sentence:

$$40\% \textit{ of } 120 \text{ miles is } 48 \text{ miles}$$
$$40\% \times 120 = 48$$

If we have three out of the four terms in a **proportion** we can always solve for the fourth by cross multiplying (this is referred to as the "rule of three"). Remember that a percentage is a ratio with the second quantity scaled to 100, this second quantity is called the "base". Given any percentage, you should always be able to answer: "Percentage of what?"; knowing that "what" is the base.

Now assume you go on a bus trip and after 60 miles are told that you have traveled 40% of the distance. What is the total distance? We are being asked: 40% of what is 60 miles? Therefore the unknown "what" is the base:

$$\frac{60 \text{ miles}}{x \text{ miles}} = \frac{40}{100}$$

To solve for the **base** we use our tried and true technique of cross multiplying.

$$\frac{60 \text{ miles}}{x \text{ miles}} = \frac{40}{100}$$

The units cancel and we get:

$$60 \times 100 = 40 \cdot x$$
$$x = \frac{6,000}{40} = 150$$

So the total distance is 150 miles, 40% of this distance is 60 miles.

*As you read, jot down notes and questions. At the end of this Guided Worksheet there is a space for **Reflection** and **Monitoring Your Understanding**, where you can try and answer your questions after completing the guided activities and homework.*

Notes & Questions

PKA Practice Problems #1 – Treatment

The 2017 National Drug Threat Assessment Report gives the following statistics in the stacked bar chart.

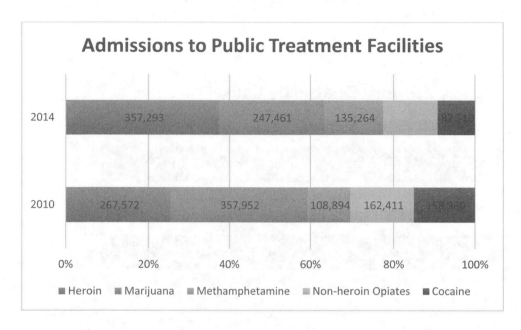

1. For 2010 compute the total number of admissions.

2. Compute the percentage of total admissions in 2010 for heroin (1 decimal place).

3. In 2014 heroin admissions accounted for 37.4% of total admissions. Compute the total (round to whole number).

4. In 2014 non-heroin opiate admissions were 13.4% of the total. Compute the number on non-heroin opiate admissions (round to whole number).

5. If the number of admissions in 2014 is less than the number in 2010 why are the bars the same length?

6. You may be tempted to conclude that since the admissions to public treatment facilities have declined from 2010 to 2014 that the number of drug overdose deaths has also gone down. Unfortunately that is not the case as shown in the following graphic. Why do think admissions are down even though overdoses are up?

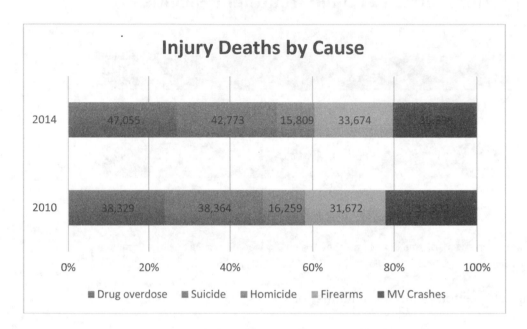

Injury Deaths by Cause

2014: 47,055 | 42,773 | 15,809 | 33,674

2010: 38,329 | 38,364 | 16,259 | 31,672

0%   20%   40%   60%   80%   100%

■ Drug overdose ■ Suicide ■ Homicide ■ Firearms ■ MV Crashes

7. Compute the percentage that drug overdose deaths are of the total for each year (1 decimal place).

8. In 2010 there were 3,007 overdose deaths due to synthetic opioids. Compute the percentage of total drug overdose deaths were due to synthetic opioids (1 decimal place).

9. In 2014 this percentage was 11.8%. Compute the number of drug overdose deaths in 2014 due to synthetic opioids (whole number).

PKA Objective 4.1 Reflections
Monitor Your Understanding

What was something you learned in this section that will make you think differently about your understanding of percentages?

FORMULA GLOSSARY

Chapter 1

Financial Functions

- Periodic rate $= \dfrac{APR}{n}$ where n = number of periods in one year

- Payment for a loan

$$PMT = \frac{P \times \dfrac{APR}{n}}{\left(1 - \left(1 + \dfrac{APR}{n}\right)^{-nt}\right)}$$

- Interest = Balance * Periodic rate
- New Balance = Balance + Interest – Payment
- APY(of a loan) = (Total interest charged for one year)/(Balance at start of year)

Descriptive Statistics

	Count	Mean	Standard Deviation
Sample Statistics	n	$\overline{X} = \dfrac{\sum x_i}{n}$	$s = \sqrt{\dfrac{\sum \left(x_i - \overline{X}\right)^2}{n-1}}$
Populations Parameters	N	$\mu = \dfrac{\sum x_i}{N}$	$\sigma = \sqrt{\dfrac{\sum \left(x_i - \mu\right)^2}{N}}$

- Mean = **AVERAGE**(numbers)
- Median = **MEDIAN**(numbers)
- Mode = **MODE**(numbers)
- Maximum = **MAX**(numbers)
- Minimum = **MIN**(numbers)
- Standard Deviation Sample = **STDEV.S**(numbers) =**STDEV**(numbers)
- Standard Deviation Population = **STDEV.P**(numbers)
- Count = **COUNT**(numbers)
- Count if a single criterion is satisfied =**COUNTIF**(range, criterion)
- Count if multiple criteria are satisfied =**COUNTIFS**(range1, criterion1, [range2], [criterion2], …)

Chapter 2

Weighted Average

- Weighted Average =

$$\frac{\sum \text{weight}_k \cdot x_k}{\sum \text{weight}_k}$$

- Weighted average using frequencies:

$$\frac{\sum w_k \cdot x_k}{\sum w_k} = \frac{\sum f_k \cdot x_k}{\sum f_k}$$

- Summing products of two columns of data = **SUMPRODUCT**(column 1, column 2)

Proportionality

	Directly Proportional	**Inversely Proportional**
Definition	Quantity A is directly proportional to quantity B if their ratio is constant: $\frac{a}{b} = k$.	Quantity A is inversely proportional to quantity B if their product is constant: $a \cdot b = k$.
Equation	$A = k \cdot B$ or $y = k \cdot x$	$A = \dfrac{k}{B}$ or $y = \dfrac{k}{x}$

Financial Literacy

- PE ratio = (share price)/(earnings per share)

Statistics

- z-scores:

$$z = \frac{x - \overline{X}}{s} \quad \text{or} \quad z = \frac{x - \mu}{\sigma}$$

- Chebychev's Theorem: the proportion of data values (for any distribution) within k standard deviations of the mean is at least:

$$1 - \frac{1}{k^2}$$

Chapter 3

Concentration and Density

- $\text{Concentration} = \dfrac{(\text{weight of powder})}{(\text{volume of liquid})}$

  ○ $= \dfrac{(\text{weight of medicine})}{(\text{weight of paste})}$

- Density = (weight of material)/(volume it occupies)

  ○ $\text{Crowd density} = \dfrac{(\text{size of crowd})}{(\text{area crowd occupies})}$

  ○ $\text{Population density} = \dfrac{(\text{size of population})}{(\text{area population lives in})}$

- $\text{Per capita} = \dfrac{(\text{amount of quantity})}{(\text{population size})}$

Statistics

- 95% rule:

$$\mu - 2 \cdot \sigma \leq 95\% \text{ of data values} \leq \mu + 2 \cdot \sigma$$

$$\overline{X} - 2 \cdot s \leq 95\% \text{ of data values} \leq \overline{X} + 2 \cdot s$$

- Standard error of the mean:

$$SE \approx \frac{s}{\sqrt{n}}$$

$$\mu - 2 \cdot SE \leq 95\% \text{ of sample means} \leq \mu + 2 \cdot SE$$

Chapter 4

Percent and Total Change/Difference

- Total change = New − Original

- Percent change = $\dfrac{(\text{New} - \text{Original})}{(\text{Original})}$

- Total difference = Quantity#1 − Quantity #2

- Percent difference = $\dfrac{(\text{Qty\#1} - \text{Qty\#2})}{(\text{Qty\#2})}$

- Growth factor = $1 + x\%$ (where $x\%$ is the growth rate)
- Decay factor = $1 - x\%$ (where $x\%$ is the decay rate)
- Percent change:
    - New = Original $\pm\, x\%$ Original
    - New = Original $(1 \pm x\%)$

Statistics

- IQR = Q3 − Q1   (interquartile range for a boxplot)

Chapter 5

Linear Functions

- Slope $(x_2 \neq x_1)$:

$$m = \frac{y_2 - y_1}{x_2 - x_1} = \frac{\Delta y}{\Delta x} = \frac{\text{Rise}}{\text{Run}}$$

- Slope-intercept equation:

$$y = m \cdot x + b$$

Statistics: Linear Regression

- Slope =**SLOPE**(y-values, x-values)
- y-intercept =**INTERCEPT**(y-values, x-values)
- Forecast =**FORECAST**(x, y-values, x-values)
- Trend =**TREND**(known y-values, known x-values, new x-values)
  - **TREND** is an array function, see text for use

Chapter 6

Exponential Functions

- Exponential equations (b = base, r = growth/decay rate, k = continuous rate):

$$y = a \cdot b^x$$
$$P = P_0 \cdot (1+r)^t$$
$$y = a \cdot e^{k \cdot x}$$

- $e^x = \textbf{EXP}(x)$

- $e^k = 1 + r$

- Factor $= 1 + r$, or $r =$ Factor $- 1$

- Given n numbers $a_1, a_2, \cdots, a_n$:

    o The **arithmetic mean** of n numbers (common usage of average value) is:

    $$\frac{a_1 + a_2 + \cdots + a_n}{n}$$

    o The **geometric mean** of n numbers (used for growth factors to find average percent change) is:

    $$\sqrt[n]{a_1 \cdot a_2 \cdot \cdots \cdot a_n}$$

    o The **harmonic mean** of n numbers (used for average rates) is:

    $$\frac{1}{\dfrac{1}{a_1} + \dfrac{1}{a_2} + \cdots + \dfrac{1}{a_n}}$$

- Given any two (input, output) data points (x_1, y_1) and (x_2, y_2) we can compute two types of average change:

    o The **average total change** is given by the formula:

    $$r = \frac{y_2 - y_1}{x_2 - x_1}$$

    o The **average percent change** is given by the formula:

    $$r = \left(\frac{y_2 -}{y_1} \right)^{\left(\frac{1}{x_2 - x_1} \right)} - 1$$

    o $1 + r = \sqrt[n]{(1+r_1) \cdot (1+r_2) \cdots (1+r_n)}$

    o $1 + r = \sqrt[n]{\dfrac{\text{New}}{\text{Original}}} = \left(\dfrac{\text{New}}{\text{Original}} \right)^{1/n}$

Statistics: Regression

- $1+r = $ **LOGEST**(y-values, x-values)
  - **LOGEST** is an array function, see text for use

- $1+r = $ **INDEX(LOGEST**(y-values, x-values), 1)
- $P_0 = $ **INDEX(LOGEST**(y-values, x-values), 2)
- Trend =**GROWTH**(known y-values, known x-values, new x-values)
  - **GROWTH** is an array function, see text for use

Chapter 7

Logarithms

- Any base b:

$$y = \log_b x \quad \longleftrightarrow \quad x = b^y$$

is equivalent to

- Common logarithms base 10:

$$y = \log x \quad \longleftrightarrow \quad y = \log_{10} x$$

is equivalent to

- Natural logarithms base e:

$$y = \ln x \quad \longleftrightarrow \quad y = \log_e x$$

is equivalent to

Properties of logarithms

- The logarithm is the power:

$$\log_b b^k = k$$

- The logarithm of a product is the sum of the logs:

$$\log_b (a \cdot c) = \log_b a + \log_b c$$

- The logarithm of a quotient is the difference of the logs:

$$\log_b \left(\frac{a}{c} \right) = \log_b a - \log_b c$$

- The logarithm of a base raised to a power is the power times the log of the base:

$$\log_b a^k = k \cdot \log_b a$$

Doubling times and half-lives

- Given a quantity that grows or decays exponentially, $P = P_0 \cdot (1 + r)^t$, the length of time before the quantity doubles or is cut in half is called the **doubling time** $(r > 0)$ or **half-life** $(r < 0)$ respectively.

  o *Doubling time*:

  $$t = \frac{\log 2}{\log(1 + r)}$$

  o *Half-life*:

  $$t = \frac{\log \left(\dfrac{1}{2} \right)}{\log(1 + r)}$$

Compounding Periodically

- Periodic compounding equation:

$$P = P_0 \cdot \left(1 + \frac{APR}{n}\right)^{n \cdot t}$$

- Annual percentage yield:

$$APY = \left(1 + \frac{APR}{n}\right)^{n} - 1$$

- Euler's number:

$$e = \lim_{N \to \infty} \left(\frac{1}{N}\right)^{N} = 2.7182818459\ldots$$

Log Plots

Given an exponential equation,

$$y = a \cdot b^{x}$$

- Taking the logarithm of both sides gives $\ln(y)$ as a linear function of x:

$$\ln(y) = \ln(b) \cdot x + \ln(a)$$

- Plotting $\ln(y)$ as a linear function of x is a semi-log plot.

Given an power equation,

$$y = k \cdot x^{-a}$$

- Taking the logarithm of both sides gives $\ln(y)$ as a linear function of $\ln(x)$:

$$\ln(y) = a \cdot \ln(x) + \ln(k)$$

- Plotting $\ln(y)$ as a linear function of $\ln(x)$ is a log-log plot.

Chapter 8

Statistics

- The **variance** for sample univariate data is the standard deviation squared:

$$\text{Var} = \frac{\sum\left(x_i - \overline{X}\right)^2}{N-1} = s^2$$

- The **covariance** for sample bivariate data is a measure of how correlated the two variables are and is given by an equation similar to the variance,

$$\text{Covar} = \frac{\sum\left(x_i - \overline{X}\right)\cdot\left(y_i - \overline{Y}\right)}{N-1}$$

- The **correlation coefficient** is defined by the formula:

$$R = \frac{\text{Covar}}{s_X \cdot s_Y}$$

- $R = $ **CORREL**(y-values, x-values)
- Covar $= $ **COVARIANCE.S**(y-values, x-values)
- For bivariate data the z-score for a y-value is proportional to the z-score for the associated x-value:

$$z_y = R \cdot z_x$$

Chapter 9

Average Percent Change

- Factor $= 1 + r$
- n = number of periods
- $1 + r = \sqrt[n]{(1 + r_1) \cdot (1 + r_2) \cdots (1 + r_n)}$
- $1 + r = \sqrt[n]{\dfrac{\text{New}}{\text{Original}}} = \left(\dfrac{\text{New}}{\text{Original}}\right)^{1/n}$
- $r = \sqrt[n]{\dfrac{\text{New}}{\text{Original}}} - 1 = \left(\dfrac{\text{New}}{\text{Original}}\right)^{1/n} - 1$

Exponential Growth

- $P = P_0 \cdot (1 + r)^t$
- $y = a \cdot e^{k \cdot x}$
- $e^k = 1 + r$
- $P = P_0 \cdot \left(1 + \dfrac{\text{APR}}{n}\right)^{n \cdot t}$
- $1 + \text{APR} = \left(1 + \dfrac{\text{APR}}{n}\right)^n$
- $\text{APY} = \left(1 + \dfrac{\text{APR}}{n}\right)^n - 1$

Excel Financial Functions

- $\text{FV} = \text{PV} * (1 + \text{RATE})^{\text{NPER}}$
- $\text{RATE} = \text{periodic rate} = \dfrac{\text{APR}}{(\text{number of periods in one year})}$
- Future Value =**FV**(RATE, NPER, PMT, PV)
- Present Value =**PV**(RATE, NPER, PMT, FV)
- Periodic Payment =**PMT**(RATE, NPER, PV, FV)
- Number of Periods =**NPER**(RATE, PMT, PV, FV)
- Periodic Rate =**RATE**(NPER, PMT, PV, FV)

Bonds

- Total Payout = Face Value + APR * Face Value * Term(years)

Stocks

- Market Capitalization = (Share Price) * (Number of Shares)

- $\text{EPS} = \dfrac{\text{Profit}}{\text{(Number of Shares)}}$

- $\text{PE} = \dfrac{\text{Share Price}}{\text{(EPS)}}$

- $\text{Dividend per Share} = \dfrac{\text{(Total Dividends)}}{\text{(Number of Shares)}}$

- $\text{Dividend Yield} = \dfrac{\text{(Dividend per Share)}}{\text{(Share Price)}}$

Chapter 10

Logic Functions
- **=IF**(logical test, value_if_TRUE, value_if_FALSE)
  - *Logical test* is an expression that is true or false using the logic operators: $=, <, <=, >, >=$, and $<>$ (does not equal).
  - *Value_if_true* will be the output if the logical test is true (this may be another built-in function like **SUM** or even another **IF** function).
  - *Value_if_false* will be the output if the logical test is false (this may be another built-in function like **SUM** or even another **IF** function).

- **=VLOOKUP**(lookup_value, table_array, col_index_num, [range_lookup])
  - *Lookup_value* is the value Excel will look for in the first column of the *table_array*.
  - *Table_array* is a table with multiple columns. Excel will look for the *lookup_value* in the first column, this first column consists of left endpoints of intervals (or bins). When Excel establishes which interval the *lookup_value* is in, it also establishes the row in the table from which the output will be chosen.
  - *Col_index_num* is the number of the column in the array from which the output will be chosen.
  - *Range_lookup* is an optional argument, entering FALSE for this argument causes Excel to look for an exact match in column one (not just the interval).

Random Functions
- **=RAND**()
  - This will generate a random number between 0 and 1.

- **=RANDBETWEEN**(bottom, top)
  - This will generate a random whole number between the *bottom* and *top* values (including the *bottom* and *top* as possibilities).

Chapter 11

Probability

- $P(\text{Event}) = \dfrac{(\text{Number of ways event can occur})}{(\text{Number of total possible outcomes})}$

  - $0 \leq P(Event) \leq 1$
  - $P(Event) = 0$ implies it will not occur, zero chance.
  - $P(Event) = 1$ implies it will occur, guaranteed 100% certain.

Fundamental Principle of Counting

- Given a series of k choices to be made, each with a different number of options: $n_1, n_2, \ldots, n_k$, then the total number of possible ways to choose 1 option from each of the k choices is the product:

$$n_1 \times n_2 \times \cdots \times n_k$$

- A **permutation** is a selection of r items from a group of n, with no item being selected twice and the order items are selected is counted differently (*abc* is a different permutation than *bca*). The total number of permutations is given by:

$$_nP_r = n \times (n-1) \times (n-2) \times \cdots \times (n-r+1) = \frac{n!}{(n-r)!}$$

- A **combination** is a selection of r items from a group of n, with no item being selected twice and the order items are selected is NOT counted differently (*abc* is the same combination as *bca*). The total number of combinations is given by:

$$_nC_r = \binom{n}{r} = \frac{_nP_r}{r!} = \frac{n!}{(n-r)! \cdot r}$$

Conditional Probability

- $P(A|B) = \dfrac{(\text{Number of outcomes occurring in both } A \text{ and } B)}{(\text{Number of total possible outcomes in } B)}$

$$P(A|B) = \frac{P(A \text{ AND } B)}{P(B)} = \frac{P(A \cap B)}{P(B)}$$

First Law of Probability

- The probability of event A and event B both occurring is:

$$P(A \text{ AND } B) = P(A) \cdot P(B|A)$$

- If A and B are *independent* events then $P(B|A) = P(B)$ and the formula is:

$$P(A \text{ AND } B) = P(A) \cdot P(B)$$

Second Law of Probability

- The probability of event A or event B occurring is:

$$\mathbf{P}(A \text{ OR } B) = \mathbf{P}(A) + \mathbf{P}(B) - \mathbf{P}(A \text{ AND } B)$$

$$\mathbf{P}(A \cup B) = \mathbf{P}(A) + \mathbf{P}(B) - \mathbf{P}(A \cap B)$$

- If A and B are *disjoint* events then $\mathbf{P}(A \text{ AND } B) = \mathbf{0}$ and the formula is:

$$\mathbf{P}(A \text{ OR } B) = \mathbf{P}(A) + \mathbf{P}(B)$$

Third Law of Probability

- The probability of event A is one minus the probability of its complement, **NOT** A:

$$\mathbf{P}(A) = 1 - \mathbf{P}(\text{NOT } A)$$

$$\mathbf{P}(A \text{ OR } B) = 1 - \mathbf{P}(\text{NOT } A \text{ AND } B)$$

- For *independent* events we can apply the **1ˢᵗ Law** to the last formula and get the *Golden Rule of Probability*:

$$\mathbf{P}(\textit{At Least One Independent Event } (A, B, C, \ldots) \textit{ Occurring}) =$$

$$1 - \mathbf{P}(\text{NOT } A) \times \mathbf{P}(\text{NOT } B) \times \mathbf{P}(\text{NOT } C) \times \ldots$$

Bayes' Formula

- The probability of a hypothesis, H_i, being the cause of an observed or given effect D:

$$\mathbf{P}(H_i | D) = \frac{\mathbf{P}(H_i) \cdot \mathbf{P}(D | H_i)}{\sum_i \mathbf{P}(D | H_i) \cdot \mathbf{P}(H_i)}$$

Chapter 12

Descriptive Statistics

	Count	Mean	Standard Deviation
Sample Statistics	n	$\overline{X} = \dfrac{\sum x_i}{n}$	$s = \sqrt{\dfrac{\sum\left(x_i - \overline{X}\right)^2}{n-1}}$
Populations Parameters	N	$\mu = \dfrac{\sum x_i}{N}$	$\sigma = \sqrt{\dfrac{\sum\left(x_i - \mu\right)^2}{N}}$

- Mean = **AVERAGE**(numbers)
- Median = **MEDIAN**(numbers)
- Mode = **MODE**(numbers)
- Maximum = **MAX**(numbers)
- Minimum = **MIN**(numbers)
- Standard Deviation Sample = **STDEV.S**(numbers) =**STDEV**(numbers)
- Standard Deviation Population = **STDEV.P**(numbers)
- Count = **COUNT**(numbers)
- Count if a single criterion is satisfied =**COUNTIF**(range, criterion)
- Count if multiple criteria are satisfied =**COUNTIFS**(range1, criterion1, [range2], [criterion2], …)

Frequency Distributions
- The **mean** for a sample can be approximated using a frequency table and midpoints of intervals:

$$\overline{X} = \frac{\sum x_i}{n} \cong \frac{\sum m_i \cdot f_i}{n}$$

Probability Distributions
- The **mean** for a probability distribution is given by:

$$\mu = \frac{\sum x_i}{N} = \frac{\sum x_k \cdot f_k}{N} = \sum x_k \cdot P_k$$

- The **standard deviation** for a probability distribution is given by:

$$\sigma = \frac{\sum\left(x_i - \mu\right)^2}{N} = \sqrt{\frac{\sum\left(x_k - \mu\right)^2 \cdot f_k}{N}} = \sqrt{\sum\left(x_k - \mu\right)^2 \cdot P_k}$$

- The **expected value** of a probability distribution is the mean:

$$\mu = \sum x_k \cdot P_k$$

Binomial Distributions

- A **binomial experiment** consists of a number (n) of trials with two possible outcomes (success or failure). The trials must be independent. The probability of success (p) and probability of failure ($q = 1 - p$) are constant throughout. The number of successes (x) form a binomial distribution, and is approximately normal when $n \cdot p \geq 5$ and $n \cdot q \geq 5$ with:

$$\mu = n \cdot p$$
$$\sigma = \sqrt{n \cdot p \cdot q}$$

- A **binomial probability** is the probability of getting x successes in n trials with two possible outcomes (success or failure). The trials must be independent. The probability of success (p) and probability of failure ($q = 1 - p$) are constant throughout. The number of successes (x) form a binomial distribution.

$$P(x \text{ Successes}) = {}_nC_x \cdot (p)^x \cdot (q)^{n-x}$$

$$P(x \text{ Successes}) = \mathbf{COMBIN}(n, x) \cdot (p)^x \cdot (q)^{n-x}$$

- The probability of at most or exactly a given number of successes in a binomial experiment:
  - =**BINOM.DIST**(number_s, trials, probability_s, cumulative)
    - *Number_s* = number of successes
    - *Trials* = number of trials in the binomial experiment
    - *Probability_s* = probability of a success in the binomial experiment
    - *Cumulative*:
      - TRUE = 1 will give the probability that there are at most *number_s* of successes.
      - FALSE = 0 will give the probability that there are exactly *number_s* of successes.

- The smallest value for which the cumulative binomial distribution is greater than or equal to a criterion value:
  - =**BINOM.INV**(trials, probability_s, alpha)
    - *Trials* = number of trials in the binomial experiment
    - *Probability_s* = probability of a success in the binomial experiment
    - *Alpha* = criterion value (cumulative probability)

Normal Distributions

- The probability of any data value being at most or exactly a given data value in a normal distribution:
  - =**NORM.DIST**(x, mean, standard_dev, cumulative)
    - x = data value
    - *Mean* = mean of the normal distribution
    - *Stadard_dev* = standard deviation of the normal distribution

- **Cumulative:**
  - TRUE = 1 will give the probability any data value being at most x.
  - FALSE = 0 will give the probability of any data value being exactly x.

- The smallest value for which the cumulative normal distribution is greater than or equal to a given probability or percentile:
  - =**NORM.INV**(probability, mean, standard_dev)
    - *Probaility* = given cumulative probability or percentile
    - *means* = probability of a success in the binomial experiment
    - *Alpha* = criterion value (cumulative probability)

- The probability of any data value being at most or exactly a given z-score in the standard normal distribution (mean = 0 and standard deviation = 1):
  - =**NORM.S.DIST**(z, cumulative)
    - z = z-score
    - *Cumulative:*
      - TRUE = 1 will give the probability any data value being at most z.
      - FALSE = 0 will give the probability of any data value being exactly z.

- The smallest z-score for which the cumulative standard normal distribution is greater than or equal to a given probability or percentile:
  - =**NORM.S.INV**(probability)
    - *Probaility* = given cumulative probability or percentile

Central Limit Theorem
- The **Central Limit Theorem** tells us that the distribution of sample means has the same center as the original population data but is less spread out:

$$\mu_{\overline{X}} = \mu \qquad \sigma_{\overline{X}} = \frac{\sigma}{\sqrt{n}}$$

- The standard error:

$$\sigma_{\overline{X}} = \frac{\sigma}{\sqrt{n}}$$

Confidence Intervals

- A **confidence interval for the mean** is a range of values used to estimate the true population mean for a given level of confidence. For a normal distribution:

$$\overline{X} - z_c \cdot \frac{\sigma}{\sqrt{n}} \leq \mu \leq \overline{X} + z_c \cdot \frac{\sigma}{\sqrt{n}}$$

Where z_c is the critical z-value used to compute the cutoffs for the interval.

- The margin of error is given by:

$$z_c \cdot \frac{\sigma}{\sqrt{n}}$$

t-distributions

- If σ is unknown and assuming that our **distribution of sample means** is approximately **normal** (either $n \geq 30$ or original population is normal), then t-scores will follow a **t-distribution** with $n - 1$ **degrees of freedom**:

$$t = \frac{\overline{X} - \mu}{\frac{s}{\sqrt{n}}}$$

- The probability of any data value being at most or exactly a given data value in a t-distribution:
  - =**T.DIST**(x, deg_freedom, cumulative)
    - x = data value
    - *deg_freedom* = degrees of freedom ($n - 1$)
    - *Cumulative*:
      - TRUE = 1 will give the probability any data value being at most x.
      - FALSE = 0 will give the probability of any data value being exactly x.

Distribution of sample proportions

- A number (n) of trials with two possible outcomes (success or failure). The trials must be independent. The probability of success (p) and probability of failure ($q = 1 - p$) are constant throughout. The proportion of successes $\left(\hat{p} = \frac{x}{n} \right)$ form a binomial distribution, and is approximately normal when $n \cdot p \geq 5$ and $n \cdot q \geq 5$ with:

$$\mu_{\hat{p}} = \frac{n \cdot p}{n} = p$$

$$\sigma_{\hat{p}} = \frac{\sqrt{n \cdot p \cdot q}}{n} = \sqrt{\frac{p \cdot q}{n}}$$

p-values

- A probability of obtaining a sample statistic that far or farther from the mean. If $p < \alpha$ reject the null hypothesis.

- NOTE: **BINOM.DIST**, **NORM.DIST**, **NORM.S.DIST**, and **T.DIST** all return probability values, and can be used to compute *p*-values!

t-test

- The *p*-value for a two-sample *t*-test can be computed using:
  - =**T.TEST**(array1, array2, tails, type)
    - *Array1* = data values for first variable
    - *Array2* = data values for second variable
    - *Tails* = 1 or 2 sided
    - *Type* = paired test, two-sample equal variances, two-sample unequal variances

Chi-squared Test

- The *p*-value for a chi-squared test for independence can be computed using:
  - =**CHISQ.TEST**(observed values, expected values)

Formula Glossary
Prior Knowledge Activation (PKA)

Chapter 2 ◆ Chapter 3 ◆ Chapter 4

Chapter 2

Fractions

- Equivalent fractions

$$\frac{a}{b} = \frac{a \cdot n}{b \cdot n}$$

Chapter 3

Fraction Arithmetic

- Addition/subtraction:

$$\frac{a}{n} \pm \frac{b}{n} = \frac{a \pm b}{n}$$

- Multiplication:

$$\frac{a}{b} \times \frac{c}{d} = \frac{a \times c}{b \times d}$$

- Division:

$$\frac{a}{b} \div \frac{c}{d} = \frac{a}{b} \times \frac{d}{c}$$

Chapter 4

Percentages

- Percent proportions:

$$\frac{\text{part}}{\text{whole}} = \frac{\text{percent}}{100}$$